Celebrating Nature by Faith

Celebrating Nature by Faith

*Studies in Reformation Theology
in an Era of Global Emergency*

H. PAUL SANTMIRE

CASCADE *Books* • Eugene, Oregon

CELEBRATING NATURE BY FAITH
Studies in Reformation Theology in an Era of Global Emergency

Copyright © 2020 H. Paul Santmire. All rights reserved. Except for brief quotations in critical publications or reviews, no part of this book may be reproduced in any manner without prior written permission from the publisher. Write: Permissions, Wipf and Stock Publishers, 199 W. 8th Ave., Suite 3, Eugene, OR 97401.

Cascade Books
An Imprint of Wipf and Stock Publishers
199 W. 8th Ave., Suite 3
Eugene, OR 97401

www.wipfandstock.com

PAPERBACK ISBN: 978-1-5326-9971-9
HARDCOVER ISBN: 978-1-5326-9972-6
EBOOK ISBN: 978-1-5326-9973-3

Cataloguing-in-Publication data:

Names: Santmire, H. Paul, author.

Title: Celebrating nature by faith : studies in Reformation theology in an era of global emergency / H. Paul Santmire.

Description: Eugene, OR: Cascade Books, 2020. | Includes bibliographical references and index.

Identifiers: ISBN: 978-1-5326-9971-9 (paperback). | ISBN: 978-1-5326-9972-6 (hardcover). | ISBN: 978-1-5326-9973-3 (ebook).

Subjects: LCSH: Ecotheology. | Nature—Religious aspects—Christianity. | Human ecology—Religious aspects. | Luther, Martin, 1483–1546.

Classification: BT695.5 S320 2020 (print). | BT695.5 (epub).

Manufactured in the U.S.A.

To David M. Rhoads
and
Lutherans Restoring Creation,
with gratitude

Contents

Preface: One Step Backward | ix

1. Living with Nature according to the Bible:
   From Stewardship to Partnership | 1

2. Martin Luther's Theology of Nature:
   Announcing the God Who Is In, With, and Under All Things | 42

3. Joseph Sittler's Pioneering Vision of the Cosmic Christ:
   Nature Transfigured | 78

4. The Theology of Nature as an Emergent Field of Promise:
   Multidisciplinary Reformation Explorations | 103

5. Celebrating Nature by Faith:
   A First-Person Theological Narrative | 131

Bibliography | 159

Index | 169

Preface
One Step Backward

THE HUMAN SPECIES, AS just about any thoughtful soul who keeps up with the daily news will have recognized by now, is currently facing a global environmental and justice emergency of unprecedented magnitude, signaled above all by the threat of human-induced climate change. Some prefer to call this the threat of climate destruction.

In recent years, leaders of the ecumenical Church have been at the forefront of those who have been calling all citizens of the Earth to address this global emergency. His Holiness, Pope Francis, may only be the most famous of those Christian leaders who have responded to the onslaught of the global crisis, with his powerful and widely hailed 2015 encyclical, *Laudato Si': On Care of Our Common Home.*[1]

With many other Christians around the globe, I, for one, as a matter of course enthusiastically welcome the current ecumenical response to our global crisis. But I also recognize that sometimes it is wise, at least for some, to take one step backward, in order to make it easier for others to take two steps forward.

1.

As a matter of fact, the Holy Father himself has already highlighted the strategy of taking one step backward in order to take two steps forward in his aforementioned encyclical and, all the more so, by his decision to take the name of St. Francis. The Pope has stepped back to reclaim the witness of that venerable thirteenth-century celebrant of nature—along with other

---

1. Pope Francis, *Laudato Si'*.

Preface

historic Catholic figures—in order to help the whole Church move forward to address the global ecojustice challenges of our own day.

Catholic theologians of the Vatican II era thought of this kind of process as a return to the sources, *ad fontes*, a return driven by a passion to then step forward into more promising theological worlds, under the guidance of the Spirit of Truth. Pope Francis is an heir of that kind of thinking.

In this book, much more modestly, I am proposing to take one theological step backward, in behalf of the two theological steps forward that I believe that the whole ecumenical Church is being called upon to undertake today, following the example of Pope Francis. Here I propose to take the reader back—*ad fontes*—to some critically important existential places in the thought and practice of the traditions of the Protestant Reformation. My purpose is to make it possible for those who join me in these retrospective explorations, especially heirs of that sixteenth-century theological movement, but also, I would hope, heirs of the Catholic Reformation later in the same century (Council of Trent), to enrich their theological responses to the global emergency we all are facing today.

To facilitate the discussion, I want to begin with some terminological observations. Many of the key constructs I will be employing in these studies I will understand in terms of ordinary speech. Thus I will typically take "nature" to mean what we commonly think of as the natural world. The only place where I will attempt to define "nature" in specific theological terms will be in my discussion of Luther's thought, for reasons that I will explain at that point. And "ecology" I will take to refer the now widely-attested scientific approach to nature as a world of interrelated subsystems. Further, the context in which "nature" or "ecology" and related words or expressions occur will, I believe, generally disclose the nuances I have in mind for a term's meaning in each instance and thereby, hopefully, avoid any serious confusion.

2.

This book itself will unfold in the following manner. I propose to take the reader first to the heart of the Reformation experience: pondering the Scriptures anew (Chapter 1). Whatever else the Reformation slogan *sola scriptura* might have meant in the sixteenth century, today it surely means that the theological heirs of the Protestant Reformation will continue to be passionately preoccupied with the Scriptures, to be in love with them,

Preface

as it were. So I will ask, first thing: how are we humans to live with nature, according to the Scriptures?

Drawing on work of my own over several decades, I will suggest in this chapter that the time is at hand for the ecumenical Church to move beyond the immensely popular theological construct of stewardship of nature to a fresh and more comprehensive construct of partnership with nature. Why? Because this is how I believe that the Scriptures mandate us to celebrate nature by faith in these times.

This chapter will also serve another important purpose. It will outline what I take to be the main lines of a biblical understanding of nature, with a strong emphasis on the Old Testament. I will go into considerable detail along the way, highlighting the findings of biblical scholars across the ecumenical spectrum whenever possible. For heirs of the Protestant Reformation, as for many other Christians these days, such biblical findings will be of the utmost importance, since most, if not all, of us are committed to regard the Bible as the primary sourcebook of our faith.

I will then take readers to a place that will come as no surprise to many Protestants, and probably not to anyone with theological interests who has ever encountered the world of Reformation thought: to consider the theology of Martin Luther himself and his own vibrant way of celebrating nature by faith (Chapter 2). This is *ad fontes* for me—as it will be for many other Lutherans and for numerous Protestant fellow-travelers, too. This is a topic that also has preoccupied me for many years.

Luther is my "Church father." He is, as it were, my St. Athanasius or my St. Francis or my St. Thomas Aquinas or my John Calvin or my John Wesley. He is first among equals, *prima inter pares*, in my own theological world. For me, Luther not only lived and spoke in the past. He also speaks with a living voice today, for better, for worse. And I know that many others in the ecumenical community today, including Pope Francis himself, also take Luther seriously in some manner, as the recent global celebrations of the five hundredth anniversary of the Reformation showed.

Just for this reason, I will try to identify the historic voice of Luther as carefully as I can. I will attend to Luther's own way of thinking in its own setting as he himself celebrated nature by faith, because Luther is the hero of a thousand causes. And I would not want to appear as if I were somehow taking him captive to my own—urgently contemporary—cause.

I will next explore the pioneering achievement of that ecumenically driven American Lutheran ecotheologian of the mid-twentieth century,

Preface

Joseph Sittler (Chapter 3), which began publicly with his address to the World Council of Churches Assembly in 1961. In that presentation and elsewhere in his writings, Sittler celebrated nature by faith in a variety of ways, above all by his call for renewed interest in cosmic Christology. Notwithstanding the sometimes less than enthusiastic theological responses that his proposals elicited in those days, Sittler believed that, by celebrating nature by faith the way he did, he was reclaiming the witness of the Scriptures and also the teachings of Luther in a manner that could and would forthrightly address the theological challenges of our own era of ecological crisis.

In the following chapter, I will highlight the promising contributions of several other American Lutheran theologians to the field of ecological theology as it emerged during the second half of the last century, in the aftermath of Sittler's proposals (Chapter 4). Theirs was a multidisciplinary celebration of nature by faith: theological anthropology (Philip Hefner), systematic theology (Ted Peters), Old Testament studies (Terence Fretheim), liturgical theology (Gordon Lathrop), the history of Christian thought (myself), and ecojustice ethics (Larry Rasmussen). In this chapter, I will also discuss the consonant emergence of a range of practical theological initiatives and their influence, above all the contributions of a number of outdoor ministries in the US.

The last of the five studies in this book will be a testimonial, a personal narrative of what it can mean to celebrate nature by faith as a Reformation theologian who has aspired to be heard as a member of the ecumenical Church (Chapter 5). I believe that the time has come for me, after working with the theology of nature for more than fifty years, to own up to the strengths and weaknesses, such as they might have been and such as I might be able to discern them, of my own theological labors. What, I will ask, does the work of a single Reformation ecotheologian of longstanding look like from within?

I will leave it to others to decide how helpful my own ecotheological labors over the past fifty years might have been: to ask how any theological dead-ends that have appeared might be avoided in the future or what insights, if any, might be claimed for further reflection. In this chapter, I especially will have college students, seminarians, younger theologians, and Church environmental activists of all ages in mind, across the ecumenical spectrum. I will be saying: these were my challenges and my struggles, theologically and vocationally. I hope that you will be able learn something from the way I addressed those challenges and struggles, positively or negatively or perhaps both, as you chart your own theological and vocational courses in this time of crisis.

I am well aware that in shaping the whole book the way I have done—highlighting not only the witness of the Scriptures, but also Luther's theology, the vision of a major Lutheran theologian, Joseph Sittler, the contributions of several other contemporary Lutheran theologians, and concluding with a single Lutheran ecotheologian's story, my own—that the whole project I have in mind could appear to be parochial, in the narrow sense of that word. Doesn't our world in crisis require multi-religious investigations or at least broadly based ecumenical responses? In these times, won't the very idea of the particularistic theological project that I am proposing here suffer from what some wag once called—hardening of the categories? I don't think so.

I don't dispute for a moment the fact that what our world in crisis needs now is to hear the voices of many religious traditions, along with the broadly based testimony of ecumenical Christianity. But there is surely a place for—possibly profound enrichment from—the witness of particular theological traditions, even denominational variants of those traditions. But it would take me too far afield to consider that kind of question in advance. I can only say this much here, invoking an ecological image. One tree doth not a forest make, surely. But sometimes it can be very helpful to understand the life of a single tree, precisely in order better to understand how to strengthen the wellbeing and even the flourishing of the whole forest.

3.

Each of the following five studies, I hope, will stand on its own. But I also hope that readers will discover a common theme, along the way, that unites them all. In this sense, the whole of this book is greater than the sum of the parts. I want to identify that theme at the outset. It has to do with the scope of theological reflection. It also has to do with what sets the kind of Reformation theology presupposed by these particular studies apart from a number of other major Western theological trends.

Consider these questions. *When you think theologically, what are you thinking about? What are the primary objects of your reflection?* For numerous prominent Christian theologians in the West, especially in the modern era, the answer to that question has been emphatically clear: *God*

Preface

and humanity. Perhaps the most important champion of this view was the theologian who has sometimes been thought of as the Thomas Aquinas of the Reformation tradition, Karl Barth (1886–1968).

Early in his professional life, Karl Barth published a book of lectures with the title *The Word of God and the Word of Man*.[2] That title told at least as much about what Barth's theology was to become as about any single theme in the book itself. Barth's mature theology was to be primarily reflections about God and humanity. Of course, following the Scriptures and the theological tradition, Barth did discuss nature from time to time. But when he did, it was primarily in terms of its meanings for the two other poles of theological reflection.

Barth stated emphatically in volume three of his multi-volumned *Church Dogmatics,* as a matter of fact, that there can be no substantive Christian doctrine of nature—as there must be, in his view, a substantive Christian doctrine of the human creature. Barth's mature theology was in this respect, to invoke his own terminology, *theoanthropocentric*. Call this Barth's fundamental theological paradigm. He read the Scriptures and he wrote his theology, voluminously, throughout his long and distinguished theological career, with that focus.

In this respect, Barth was, in significant ways, a representative figure. Numerous Christian theologians, beginning in the earliest era of Christian thought, likewise focused their theology, chiefly if not totally, on God and humanity.[3] This was particularly true of Christian theologians in the modern era, especially those who identified with Reformation traditions, as we shall see at various points along the way in this book. But Barth carried through his theoanthropocentric program with a self-conscious rigor that is likely unequalled in Western Christian thought.

This was a fateful development. If theology is fundamentally theoanthropocentric, then the natural world will have its ultimate meaning, its *raison d'etre*, only in terms of God and humanity, as a kind of appendix. Nature will be allotted no integrity of its own in the greater scheme of things. Nature, at best, will have instrumental meanings. This was illustrated sharply by Barth's sometime theological opponent, Emil Brunner, who, in this respect, was Barth's compatriot. "The cosmic element in the

---

2. Barth, *Word of God and the Word of Man.*

3. I have traced this history in my study of Christian attitudes toward nature: Santmire, *Travail of Nature.*

Preface

Bible," Brunner once observed, "is never anything more than the 'scenery' in which the history of mankind takes place."[4]

The studies in this book presuppose a different theological paradigm. Since I completed my doctoral dissertation on Karl Barth's theology of nature in 1966, I have explored, in a variety of ways, the promise of thinking theologically in terms of *God, humanity, and nature* together. Call this an integral view of nature. In response to Barth's theology, I have chosen to work with what I like to call, reconfiguring Barth's own terminology, a *theocosmocentric* paradigm. This way of thinking takes God's purposes with the whole natural world just as seriously as God's purposes with humanity in particular. But I was not alone, in this respect, by any means. As we will see, Joseph Sittler (Chapter 3) and a range of other Reformation theologians also took the theocosmocentric paradigm for granted (Chapter 4).

I hope, then, that the following five studies will show, each in its own way, the fruitfulness of thinking theologically, most fundamentally, in terms of God, humanity, *and* nature, not just in terms of God and humanity alone—with nature then subordinated to the first two or even neglected altogether. Each of these studies illustrates the promise of thinking theologically in terms of the theocosmocentric paradigm—according to which nature has its own meanings in the greater scheme of things.

In a manner of speaking, therefore, the underlying theme of this book is that *nature itself cries out to be celebrated by faith*. This is the single most important learning that I hope that readers will carry forward with them, once they have taken the one step backward that I will be describing in the following pages.

I am well aware that the terminology here—theoanthropocentrism and theocosmocentrism—is not felicitous. But I do believe that it can nevertheless be fundamentally instructive, especially for explorations in the theology of nature, such as the ones that are to follow.

4.

Then some comments about the tonality of this book. Those with ears to hear will already have detected that the following explorations will be consistently predicated on the assumption that the human species is facing a life-or-death planetary emergency today: due to what human abuse,

---

4. Brunner, *Revelation and Reason*. 33n.

Preface

especially by the powerful, has done and continues to do to this, God's good Earth and to the poor of the Earth. Call this a tonality of emergency.

At the end of Chapter 4, indeed, I will raise the question whether this Earth-crisis, driven as it is by climate destruction perpetrated by the principalities and powers of this world, has become so extreme that the Church Catholic, in all its formations, must not now enter into what was called by resisting Christians in Germany during the Nazi era a "state of confession," a *status confessionis*.

Indeed, must not the very word *martyr*, which originally meant "witness," once again be deeply and self-sacrificially claimed for themselves by all faithful members of the Church of Jesus Christ? Has not the time arrived for all Christians, not just a few, to stand ready to put our livelihoods and even our very lives on the line?

I highlight this matter at this preliminary point so that no reader will conclude anywhere along the way that this book takes for granted some kind of human business-as-usual on planet Earth today. It does not. On the other hand, I have long been convinced that we who think of ourselves as Christians, together, must always try to do everything at once.

But all cannot stand in the halls of government or out on the streets all the time. Some must take their stands in those places, for sure, but others are needed to search the Scriptures, for example, or to work at constantly renewing the Liturgy, for the sake of more resolutely celebrating the presence of the God of hope in these times of pervasive ecological crisis and widespread cultural despair.

To invoke a few traditional liturgical words, which were familiar to me in my earliest years: it is "truly meet, right, and salutary" for those of us who think of ourselves as Christians today, and for the seekers in our midst as well, to write or to read books like this, as long as we understand that we are living in an era of global emergency and as long as all of us are always poised to take public stands at any time or at any place in behalf of the Earth's future and in behalf of the poor of the Earth.

With that sense of urgency, I look forward to conversing both with scholarly and general readers in this book. For the former I have provided notes so that you can more explicitly understand the theological milieu that informs this book, as you ponder your own theological pilgrimage. For the latter, feel free to skip the footnotes and then to dream dreams and see visions immediately, each step along the way, about how you yourself might enrich your own celebration of nature by faith.

Preface

5.

I enthusiastically dedicate this book to all the ecojustice activists associated with Lutherans Restoring Creation (LRC)—online at LutheransRestoringCreation.org/—who are hard at work in concert with their peers in the ecumenical community in behalf of an ecological reformation of Christianity at the grassroots, where this urgent challenge may matter the most. In the same breath, this book is also dedicated to Dr. David M. Rhoads, Professor of New Testament emeritus at the Lutheran School of Theology in Chicago and founder of LRC. David has been the heart of theologically informed American Lutheran ecojustice initiatives for decades, and for that, as well as for his personal friendship, I am most grateful.

Finally, I want to thank the distinguished Finnish American artist Eric Aho for permitting me to reproduce his striking gouache and watercolor painting as the cover of this book. For me, Aho's creativity discloses both the dark ambiguity and the luminous promise of nature charged by the presence and the power and the mystery of God. I take that dark ambiguity with utmost seriousness, especially in these times. As this book was going to press, all hell was breaking loose. The chaotic epoch of the corona virus was upon us all. But the eruption of that plague of nature just happened to coincide, in my own world, with celebrations of Easter, in whose light I also see the luminous promise of nature, notwithstanding all the darkness. As I contemplate Aho's painting, therefore, I am overwhelmed by this contradictory but captivating vision of nature, which all the explorations of this book presuppose: the struggles between the powers of death and the powers of life.

H. P. S.
Easter 2020

1

Living with Nature according to the Bible
From Stewardship to Partnership[1]

How shall we humans live with nature? That is the chief question I want to have before us in this chapter. And that, I take it, is one of the premiere questions facing the human species today.

There are other premiere questions familiar to us, to be sure. Such as: how shall we humans live with each other on planet Earth? And: how shall the poor and oppressed human creatures and indeed all the living creatures of the Earth secure their own places so that they can justly flourish? All these questions, of course, are interrelated and none of them can be rightfully answered apart from the others.

But I want to focus on the first question here, about how we humans are to live with nature, as long as it is understood that the other two are very much on my mind, too. As, moreover, those of us who identify with the Christian tradition today are driven to ponder that first question, many of us, especially those of us who treasure Reformation traditions, will as a matter of course also find ourselves driven back to the Scriptures.

So moved, we may well continue to ask Martin Luther's own premiere question as he searched the Scriptures—how shall I find a gracious God? But, with no less spiritual passion than Luther's, when he asked his question of the Bible, we will surely want to ask the question I have already highlighted—how shall we humans live with nature? Or, to use a term with more biblical resonance: how shall we humans live with "the earth"?

---

1. This chapter is a thoroughly revised version of Santmire, "Partnership with Nature," which first appeared in the *Christian Scholar's Review*.

That question has, of course, preoccupied concerned citizens on planet Earth with a sense of urgency in recent decades—although, along the way, other concerns like the pursuit of power or the pursuit of happiness or the pursuit of security appear to have pushed that urgent question out of sight and therefore out of mind, all too often. But for those Christians who have refused to be distracted by those other popular concerns one answer to the question of living with nature has been offered, again and again, and on biblical grounds: we humans shall live with nature by being good stewards.

I believe that there is a better answer to that question, and, indeed, that that better answer can be abundantly illustrated as we search the Scriptures: *partnership, rather than stewardship*. I propose to make that case now, with a fresh approach to relevant biblical texts. As these explorations unfold, too, I hope to show that the partnership idea opens up—as a kind of exegetical key—wider and perhaps not yet widely acknowledged vistas of biblical apperceptions of nature that are exciting to contemplate. But first I want to give the stewardship idea its due. In a sense, it has served us well.

The Force of a Familiar Theme: Stewardship

Stewardship is a venerable theological theme. No one can legitimately fault the National Religious Partnership for the Environment, the Advertising Council, and the Environmental Defense Fund for working together to encourage religious communities and their members to respect the earth, to "reduce, reuse, recycle," to buy recycled goods, and to use energy efficiently, all for the sake of environmental justice. The rationale for this campaign seems to be fundamentally sound, too, biblically and theologically: "The earth is the Lord's. We are its stewards."[2]

Stewardship, as a matter of fact, has been promoted by American Churches with some success, more than any other theological theme, in response to the global environmental crisis. In support of that commitment, in the past few decades, an enormous amount of biblical and theological work has been invested in defining and defending the theology of stewardship, some of it highly sophisticated.[3] Of particular importance, in this

---

2. From a poster made available to parishioners in a smalltown congregation in rural Maine.

3. Probably the single most enlightening work of many good studies on stewardship is Hall, *Imagining God*.

regard, is the prophetic emphasis on "ecojustice" espoused by some of the most outspoken advocates of the stewardship theme.[4]

Along with these kinds of theological and ethical commitments, ecumenical and denominational agencies and local congregations have also joined with various other public groups in the name of better stewardship of the Earth's resources, in order to influence the policies of timber companies, agricultural conglomerates, and other corporate interests, sometimes with tangible, positive results. Call all this the first wave of theological responses to the global environmental crisis.

With this, I am well aware, as a former parish pastor and a teaching and writing theologian, that the construct of stewardship has regularly been invoked in Church circles, often enthusiastically, for another purpose, too: to undergird Church *fundraising*. Many Christian congregations in the U.S. have regular "stewardship campaigns" every fall. Such fundraising efforts are often supported by national agencies, some of them ecclesial, some of them independent, on occasion in competition with each other. In this sense, stewardship has become a major Church industry in the U.S. in our times.

Fatefully, Church fundraising campaigns that invoke the stewardship construct also typically reach out to embrace—nature. We are told by the official spokespersons for these campaigns to be thankful for all God's blessings: thankful for the blessings of the creation as well as for the blessings of our own financial resources. This is my judgment about these matters: *I believe that the word stewardship, if it is to continue to be publicly employed in Church circles, should henceforth be restricted to such financial campaigns, and not be used to describe what Christians' relationships with nature ought to be.* In a word, let Church fundraising be one thing and Church engagement with nature be another. Indeed, the construct of stewardship, when understood as the chief frame to describe normative human relationships with nature, has a number of problems. However it might be interpreted or reinterpreted, the idea of stewardship tends, by default if not by intention, to suggest an anthropocentric, managerial approach to nature. I will not argue that point here.[5]

---

4. An impressive representative of this kind of accent on ecojustice is the document produced by the Presbyterian Church (U.S.A.) some years ago, *Keeping and Healing the Creation*.

5. I have discussed my reservations about ecclesial use of the stewardship construct in Santmire, *Ritualizing Nature*, 251–57.

Instead, as I have already signaled, I want to propose a much more comprehensive construct to describe the biblical understanding of how we humans are to live with nature—partnership. The time for the Church's theologians, preachers, and lay leaders to launch this second wave of understanding the human relationship to nature has now arrived. This is all the more a theological imperative since there had been a revolution in scholarly studies of the biblical theology of nature in recent decades.[6]

---

6. Scholarly study of the biblical theology of nature, especially the Old Testament, has developed dramatically over the last thirty years, both quantitatively and qualitatively. Some of these trends of research are reflected in this essay, as the notes will indicate. For a review of these developments through 1996, see Hiebert, "Re-Imaging Nature." Perhaps the two most substantive works yet to appear in this field are Hiebert's own study, *Yahwist's Landscape*; and Brown, *Ethos of the Cosmos*. Also important and trend-setting are several articles and books (some of them cited below in Chapter 4) by Terence Fretheim, in particular his study *Suffering of God*; and Simkins's comprehensive study, *Creator & Creation*. Rolf P. Knierim has contributed to these developments, too; see Knierim, "Cosmos and History in Israel's Theology," 171–224. All of these works are to some degree dependent on the pioneering research of scholars whose most relevant essays have been gathered in Bernhard W. Anderson, ed., *Creation in the Old Testament*. Contributors to the still-unfolding Earth Bible series, edited by Norman C. Habel (Cleveland: Pilgrim, 2000 et seq.), have also advanced the discussion in recent years.

On the other hand, scholarly investigation of the theology of nature in the New Testament has not advanced the way it has in Old Testament studies. For a variety of reasons that cannot concern us here, the most interesting research has generally been confined to relatively limited topics and has appeared mainly in professional journals. The scholarly work that has been done, however, is most promising, particularly studies of the eschatological cosmology of Romans, the cosmic Christology of Colossians and Ephesians, and the nature-affirming theology of the book of Revelation. Worthy of special attention in this regard is the thorough, groundbreaking study by Edward Adams, *Constructing the World*; and, perhaps the two most important exegetical books as of this writing: Horrell, et al., *Greening Paul*; and Bauckham, *Bible and Ecology*.

The time seems to be at hand for the appearances of more extensive New Testament studies of this kind. To this point, however, with the exception of books like the three I have just mentioned, the most probing scholarly discussions of the topic have been produced mainly by theologians such as Jürgen Moltmann, who have forged ahead, perhaps necessarily so, in the absence of such works by New Testament scholars themselves. See especially Moltmann, *Way of Jesus Christ*. See also the concluding chapter of my own study—Santmire, *Travail of Nature*; the chapter called "A New Option in Biblical Interpretation," looks back at the theology of the Bible from the perspective of a fresh reading of the classical Christian tradition, and makes several heuristic proposals for reading the New Testament theology of nature, in particular (see 200–218).

The Emergence of a Biblical Theme: Partnership with Nature

This is my contention: that in order to reflect the complexities and the richness of biblical testimony in this respect, it is best for us to develop a theology of partnership with nature, which will hopefully, in due course and with sufficient scholarly discussion, begin to take the place of what appears to be the more limited theology of stewardship of nature which now is being widely preached and taught.[7]

And more. I hope that it will be apparent that this chapter's argument will also have the effect of reinforcing the theme of *theocosmocentrism* in theology, a construct which I have already highlighted. The idea of human partnership with nature presupposes that nature has its own meanings in the greater scheme of things, that the purpose of nature is by no means just to serve, much less to be a slave to, human beings. Theocosmocentrism presupposes that God has a history of God's own with the whole cosmos, from Alpha to Omega, along with humans, who themselves are embodied in the cosmos.

---

7. In this sense, founders of the National Religious Partnership for the Environment (referred to above) were pointing the way, knowingly or not, to what I am identifying as a second wave of theological responses to the environmental crisis when they chose to think of themselves in terms of the language of *partnership*. But they were very much being carried along by the first wave of responses, too, insofar as they opted to explain their purposes in the language of *stewardship*.

The language of partnership has also been employed by Norman C. Habel in the Earth Bible series, as one of the "Six Ecojustice Principles" that shape the explorations of those who contribute to the series. This surely advances the discussion. The problem with this particular usage, however, is that the language of partnership is tied to the construct of "custodianship." However much the latter term might be redefined, in an attempt to transcend the older stewardship language (as Habel and his colleagues want to do), the idea of a custodian ends up sounding very much like the idea of a steward. Which means, in effect, that the new construct of partnership is still implicitly bound, in not altogether helpful ways, to the old construct of stewardship. Cf. *Readings from the Perspective of Earth*, 24: "*The Principle of Mutual Custodianship:* Earth is a balanced and diverse domain where responsible custodians can function as partners, rather than rulers, to sustain a balanced and diverse Earth community."(italics original) *In this essay, I understand partnership to be* the *generic theological term, which shapes all others.*

Of interest, too, in this connection are some of the works of Letty M. Russell, especially Russell, *Future of Partnership*; and Russell, *Growth in Partnership*. But while she demonstrates that "partnership" is a *bona fide* theological construct, she shows little interest in the theology of nature in these studies and also, at various points, she works within the framework of a theology of stewardship.

The idea of human partnership with nature also resonates with the theme of cosmic Christology, given prominence by Joseph Sittler, which I will review in Chapter 3. Jesus Christ is, indeed, the Savior of the whole world, the cosmos (cf. John 3:16). All the creatures of nature, in this sense, are worth saving! Nature is by no means just a platform on which God's history with humans can unfold. Nor is it merely a resource that makes human life possible and which therefore can be sloughed off by God, as it were, when the eschatological new creation fully and finally arrives.

On the contrary, according to the Scriptures, we are to hope for new heavens and a new earth (2 Pet 3:13), when the lamb will lie down with the lion and a little child will lead them, when all creatures will be born anew eternally, not just human creatures. In a word, nature has its own integrity, both now and in the Age to Come. That's why humans are created by God to be partners with nature, not overlords or masters.

The biblical theology of partnership with nature, however, it will soon become apparent, is by no means a one-dimensional construct. We will encounter its complexity and richness in three fundamental expressions or emphases, under the following rubrics:

- Creative intervention in nature,
- Sensitive care for nature, and
- Awestruck contemplation of nature.

The first of these emphases—creative intervention in nature—will, I believe, validate and also in some fundamental ways correct standard theological expositions of the stewardship doctrine as it has generally invoked in the Church's theological discourse regarding nature. Overall, therefore, the theological-exegetical result of the explorations that follow will hopefully be *something much more*, rather than something much less or something totally different from what the familiar construct of stewardship of nature has offered us.

Considering the Witness of the Scriptures as a Whole

To identify biblical teachings about partnership with nature now, it will be necessary for us to frame our exegetical sights broadly. It will be critically important, at the outset, to identify the witness of the whole Bible, however immodest this kind of goal might sound at first, particularly within the parameters of a single chapter. Why? All too often, in contexts like this one, where ecological issues are at the fore, the discussion grinds to a polemical impasse because it never gets beyond proof-texting, whether of the secular or the theological variety; or, more irenically, it gets mired in word studies.

It is surely of interest, for example, what Genesis 1 has to say about human "dominion" over the earth and "subduing" the earth, but that is by no means the whole biblical story, however those particular words might be interpreted. For such reasons, I self-consciously choose to join with those Church theologians who position themselves to listen to the witness of the whole Bible as the Word of God to us, and to do so with a "critically engaged reading."[8]

This means, among other things, reading the Scriptures as the Church has traditionally read them, as telling the Great Story of the "God of grace and God of glory"(Harry Emerson Fosdick).[9] This Great Story is the revelational whole which is larger than the sum of the historical parts of Holy Scripture, although known only in the encounter with those historical parts. This is the story "that runs from creation to new creation, with the

---

8. Cf. Green, "Scripture and Theology," 5–20, esp. 19: "A critically engaged reading of scripture would account for the text in its final form; for the text as a whole; for the cultural embeddedness of all language (rather than assuming that all people everywhere and in all times construed their life-worlds as we do); for the location of particular canonical witnesses within the grand mural of the actualization of God's purpose; and for the witness of scripture as seen in its effects within the community of God's people—not least in the distillation of scripture's message in the great creeds that confess and proclaim and worship the triune God."

9. This affirmation, to be sure, contradicts the theological or metaphysical assumptions firmly held by many "post-modern" thinkers and other critics, who typically assert that there can no longer legitimately be any kind of comprehensive statement of ultimate meanings, any "meta-narrative." But from the perspective of the Church, this precisely is what we Christians have been called to do, not uncritically, but passionately: "to tell the old, old story of Jesus and His love," to proclaim the "good news" to every creature. For an insightful diagnosis of the anguish of the "modern mind" and its post-modern heir, in this respect, see the short essay by Robert Jenson, "How the World Lost Its Story." I also want to note here that I myself prefer the language of "story" in this context, because of its concreteness. "Great Story," for me, is much more expressive than other constructs, such as "metanarrative."

Christ-event as its interpretive middle."¹⁰ Wolfhardt Pannenberg has taught us to speak of that story, more conceptually, as God's *universal history*.¹¹

This is a story of the genesis and the fulfillment of all things, which begins—and in some sense is already completed, implicitly, as we shall see, with the witness to the seventh day in Gen 1—with the first chapter of the first book of the Bible, and which concludes, climactically and expansively, with the explicit eschatological witness of the last book of the Bible.¹² This

---

10. Cf. Green, "Scripture and Theology," 20: "As the instrument of revelation, scripture presents the paradigm by which Christians make sense of the world in relation to God; it is therefore incumbent on Christians to engage in theological reflection on scripture whereby their imaginations are yielded to its theological vision. This practice will not overcome the problem of diversity of perspectives within the Bible, but it will focus our imaginative faculties on the pattern of the story ('the plot') that runs from creation to new creation, with the Christ-event as its interpretive middle."

11. For one interpretation of Pannenberg's theology, see Buller, *Unity of Nature and History*. Cf. also the comment of Terence Fretheim, "Nature's Praise of God in the Psalms," 26: "God is not only a God of history; God is also a God of nature. Or, to put it in other terms, God is as active in the history of nature as God is in the history of humankind; from God's perspective we have to do with one history and one world."

12. This approach to Scripture in terms of universal history takes the place of—but does not totally supplant the content of—what biblical and systematic theologians in the last century, such as G. Ernest Wright, Gerhardt von Rad, Oscar Cullmann, Rudolf Bultmann (in existentialist terms) and Karl Barth, to mention a few, used to refer to as the history of salvation (*Heilsgeschichte*) as the fundamental biblical interpretive category. Exegetically, the history-of-salvation approach typically set salvation over against creation in general and nature in particular. Nature was viewed as the stage for the history of salvation. For someone like Wright, in particular, there was even a self-consciously affirmed Hegelian dialectic: a religion of salvation-history, that is, the faith of ancient Israel [thesis], encounters a religion of nature, that is, the faith of the Canaanite Baal [antithesis], and then expands to claim that religion of nature as a presupposition for the religion of salvation-history [synthesis]. That view still finds advocates, some of them scholars of renown, such as Childs, *Biblical Theology of the Old and New Testaments*, 110: "To summarize. Israel's faith developed historically from its initial encounter with God as redeemer from Egypt, and only secondarily from this centre was a theology of creation incorporated into its faith." Interestingly, in order to sustain this position, Childs has to qualify his own "canonical criticism," since the Bible obviously begins with—creation! He maintains that "the present canonical shape has subordinated the noetic sequence of Israel's experience of God in her redemptive history to the ontic reality of God as creator. This is to say, although Israel undoubtedly first came to know Yahweh in historical acts of redemption from Egypt, the final form of the tradition gave precedence to God's initial activity in creating the heavens and the earth" (385). It is also interesting that Childs explicitly sides with the theoanthropocentrism of Karl Barth, in this respect (386) [on Barth, see the references at the end of this note].

In recent decades, many biblical scholars, particularly Old Testament specialists, have left that history-of-salvation approach behind, primarily on the grounds that it does not

is the story of God's history with the whole creation, brought to its glorious fulfilment by the life, death, resurrection, and ascension of God's Son, Jesus Christ, all in the power of the Holy Spirit.

This is "the old, old story of Jesus and His love" (Katherine Hankey): of how God graciously resolves in eternity to share the Divine life and the Divine love with a good and beautiful world of creatures in a universal history, in the midst of which God resolves to become incarnate in Jesus Christ, in order to restore the by-then long fallen human race to its intended place of fidelity within the whole of creation-history and then to carry that whole history on to its final and glorious fulfilment in new heavens and a new earth, in which righteousness will dwell.[13] Which is also to say: as we approach the Great Story voiced by Scripture, we know who the chief actor is, *God*, and what motivates everything that this God does in God's universal history—*self-giving love*.[14]

---

adequately represent the history of Israel's (and, later, the early Christian community's) faith. *The God who was Israel's redeemer, many have argued, was from the very beginning, for Israel, the Creator of the world.* For an introduction to this discussion, see Hiebert, "Rethinking Traditional Approaches," 23–30; and Fretheim, "Reclamation of Creation." See also Anderson, "Creation and Ecology"; Schmid, "Creation, Righteousness, and Salvation"; and Landes, "Creation and Liberation," 135–51—all in in Anderson, ed., *Creation in the Old Testament*.

My own work on Barth in 1966 raised questions about the primacy of the *Heilsgeschicte* motif in dogmatic and systematic theology, in significant measure on biblical grounds. For purposes that made good sense to Barth when he wrote his monumental *Church Dogmatics*, Barth shaped his theology as theoanthropocentrism, that is, as chiefly a doctrine of God and humanity. Barth maintained, in turn, as I have already noted, that there can be no legitimate, substantive Christian "theology of nature." See my doctoral dissertation: Santmire, "Creation and Nature," an analysis that I later summarized in Santmire, *Travail of Nature*, 145–55. That Barthian theoanthropocentric focus is no longer acceptable, first and foremost, because it is not fully biblical, as the following discussion will show. That we also face challenges that differ from the ones that Barth faced is, of course, also true.

13. I described the shape of a theology predicated on this kind of understanding of the witness of Scripture in my early, programmatic study, Santmire, *Brother Earth*.

14. In his study, *Suffering of God*, 23, Terence Fretheim has rightly stressed the importance of biblical generalizations about God, confessional constructs, which, I likewise believe, the Great Story of Scripture presupposes, above all the theology of God's self-giving love or constant mercy. "Only such generalizations, irreducible to story form," Fretheim instructively argues, "enable one to discern continuities, or something strange or new, and to realize when an objective has been reached . . . Thus story and generalization must be kept inextricably interwoven; together they determine the structure and form of OT theology."

Celebrating Nature by Faith

In the midst of that revelational biblical macro-narrative, we can then see signs of several historical micro-narratives or thematic trajectories in the Scriptures, all of which enrich and help to define the whole, among them: the Priestly, the Yahwistic, the Deuteronomistic, the Prophetic, the Sapiential, the Jobean, and the Messianic/Eschatological trajectories.[15] For the purposes of this chapter, I want to give most of my attention to the Priestly, the Yahwistic, and the Jobean micro-narratives, not for a moment denying the importance of the others or of still other, complementary materials, like the Psalms.

This approach will allow us identify, if not fully describe, the biblical "narrative base line," which is important not only in itself, but also as a witness to the redeemed life. Whatever else redemption means for those of us who are Christians, it surely means this much: that by his death and resurrection *Jesus Christ restores those who believe in him to the existential location where they should have been in the first place, had they not been sinful.*[16]

---

15. This is to adopt and adapt the kind of interpretive approach used in Brown, *Ethos of the Cosmos*, who discusses these micro-narratives extensively, with the exception of the Deuteromistic and the Messianic. For explorations of the traditions of the Deuteronomic history, with particular reference to our theme, see the materials covered in Habel, *Land Is Mine*. Jürgen Moltmann has given perhaps the most compelling statement of biblical messianism, seen in a cosmic context, in Moltmann, *Way of Jesus Christ*. For specific expositions of biblical messianism in terms of cosmic Christology, see Wilkinson, "Cosmic Christology and the Christian's Role in Creation,"; and also my own reflections in Santmire, *Nature Reborn*, ch. 4.

The approach to biblical interpretation I am taking here presupposes many of the concerns that inform the work of Brevard S. Childs and the "canonical criticism" that he advocates; see especially Childs, *Biblical Theology of the Old and New Testaments*. Yet I agree with Terence Fretheim's argument in his aforementioned study, Freitheim, *Suffering of God*, 19, in favor of what I take to be a more richly dialogical hermeneutic than Childs's, a hermeneutic which hears the one Word of God speaking through the Bible in *many* voices, not just through the voice of the final redactors of the Scriptures: "[S]uch a perspective [canonical criticism] fails to recognize adequately that pluralism of OT theologies has in fact issued in a pluralism of Christian theologies, despite a finally shaped canon, just by virtue of giving greater attention to this or that aspect of the OT. This pluralism of OT interpretation is already evident in the NT, which issues in a new pluralism of its own. Even if the latest redactors of the OT interpreted the whole OT in terms of a unified perspective, the fact of theological pluralism in the NT and the church means that a truly canonical approach ought to be as concerned with theological diversity as unity, as well as being especially open to new insights which might have their roots in a heretofore undiscovered dimension of this pluralism. *Pluralism has been canonized*" (italics original).

16. I have explored this understanding of the work of Christ—as restoration—along with other motifs in Santmire, *Brother Earth*, ch. 8.

It also means that by his death, resurrection, and ascension Jesus Christ begins to carry us and all creatures forward toward the eschatological new creation, when all things will be made new and when God will be all in all.

But I want to focus here on the first, the restorative meaning of the advent of Jesus Christ. According to this reading of the Scriptures, as members of the faithful community of Christ, each of us becomes, in Christ, a new Adam or a new Eve—a new being—as we are claimed by the redemptive work of the Second Adam, Jesus Christ, whom Paul Tillich called the New Being. The witness of the Priestly writers, the Yahwist, and Job helps to identify that new place for us, which is ours, as believers, through Jesus Christ, a place where we should have been living had we not become captives of the powers of sin, by habitually reenacting the primordial scenario narrated in the primeval Adam and Eve story.

What, then, is our Divinely mandated relationship with the good creation-history of God to which God's redeeming act in Jesus Christ restores us? That is the question that will preoccupy us now, as we explore the witness of the aforementioned three micro-narratives, with a view to identifying how we humans are intended by God, biblically speaking, to live with nature in partnership.

In following the biblical narrative this way, I want to emphasize, I will be describing *dimensions* of creation-history, not literal, temporal stages or "days" of creation-history. To talk about the Garden of Eden, for example, is to talk about the city of Watertown, Massachusetts, where I currently live. It is not to talk about something that happened only way back when—"at the beginning of time." The soil with which I work when I take care of my house plants in this sense is the soil of Eden. It is good. It is created and blessed by God, likewise for the red-tailed hawks that fly by my ninth-floor window and the storm clouds and the lightning bolts and, sometimes, the gorgeous rainbows that I can see out the same window. In this sense, I live in an Edenic world.

On the other hand, this Edenic world in which I work with my hands and which I see when I look out my window could be destructively overrun by the chaos unleashed by rising seas in the next several decades, due to climate change, induced by human greed and misunderstanding and indifference. And I am both a willful and an unwillful participant those destructive global trends. So I am living outside of Eden, too. I most certainly live in a way, estranged from God and from all other creatures, which contributes to the "mass of perdition"[17] that burdens human history and indeed the whole earth and all its creatures as never before and gives some sensitive souls the dread that we may be living in the Last Times (as Martin Luther believed about his own era).

The point is this: the biblical writers, when they talk about primal beginnings, are talking about *this* world, which we all experience every day.

The Witness of Genesis 1

The whole biblical story commences with an account of the beginning of all created things and, implicitly, the continuation and the fulfilment of all things. We have this micro-narrative in Gen 1:1—2:3(4a) from the hands of traditionalists—or tradents—who are usually called the Priestly writers. My intent here is not to offer anything resembling a complete exegesis of the Genesis 1 creation narrative at this point,[18] nor to deal directly with critically important doctrinal issues, such as "creation out of nothing" *(creatio ex nihilo)*, but to highlight some of the major themes that appear in this tightly packed and carefully composed chapter, as they help us to catch sight of the biblical theology of nature in all its rich diversity. This text is, indeed, an introduction to the witness of the whole Bible, and, as part of that introduction, a witness to the rich cosmic concerns of the whole Bible, from the very beginning.[19]

The first verse of this micro-narrative speaks the most important word—*God*. However the first verse of Genesis 1 is translated—and this is much discussed—whether it be the version preferred by the NRSV, "In the beginning when God created the heavens and the earth . . ." or, as in the NRSV notes, "when God began to create . . ." or "In the beginning God

---

17 Augustine, *Enchiridion*, 107.

18. Some of the most insightful exegesis of the Genesis creation accounts is to be found in Brown, *Ethos of the Cosmos*, chaps. 2 and 3. Also, see Terence Fretheim, "Creator, Creature, and Co-Creation in Genesis 1–2"; and Santmire, "Genesis Creation Narratives Revisited."

19. Cf. Brown, *Ethos of the Cosmos*, 36: "the Priestly account of creation in Gen 1:1—2:3(4a) commands an unassailably preeminent position. This cosmic overture to the entire canon is the literary and theological point of departure for all that follows, from creation to consummation. By virtue of its placement at the Bible's threshold, this quintessential creation story not only relativizes the other biblical cosmogonies interspersed throughout the Old Testament, but also imbues all other material, from historical narrative to law, with cosmic background."

created ...," the whole point of this crucial text is *the God who creates,* and in view of the immediate story that is to follow in Genesis 1 and later testimonies in the Priestly narrative, such as the covenant of promise that God makes with all creatures in the Noah story: this is *the God who creates in order to give of Godself* so that a whole range of creatures might have being and life, and have it abundantly, in a history with God, who will be faithful to the Divine promises, come what may. Genesis 1:1 begins a story, which, however circuitous it may be at times, however interrupted it may be on occasion, obscured by experiences that point in other directions, is a story of a God "whose giving knows no ending" (Robert L. Edwards).

And this self-giving of God, according to the Priestly tradents, as in Holy Scripture generally, is always understood not as some impersonal force, however serendipitous (Gordon Kaufman) that force might be envisioned to be, but as an amazing and mysterious *personal* giving, a personal sharing, a partnering in that sense, as is indicated in Genesis 1 by the repeated witness to God *speaking*. A "force" does not speak. We will encounter this motif more than once, when we review the witness of the Yahwist and of Job, as well.

The Divine speaking, as Martin Buber showed in various ways, always signifies the Divine commitment to personal sharing, to be an I who is akin to the I of an I–Thou relationship with another and to give of oneself to the other. That kind of relationship will differ, to be sure, with different kinds of creatures, as Martin Buber already recognized as he pondered what an I–Thou relationship with a tree might or might not mean.[20] Perhaps the best, if most compact, way to express this subtle distinction between the two overall kinds of personal relationships—God in intimate personal relationship with a variety of creatures and God in intimate personal relationship with the human creature in particular—is to see all God's *ad extra* relationships as God's gracious *self-communication* to others and the particular relationship of God with the human creature as God's gracious self-communication to an other in the form of *communion*. In other words, God's communication with the human creature is more internal than external, more intangible than tangible, known more by insight than by sight.[21]

---

20. For a development of this theme, predicated on Buber's thought, in terms of a not-yet-mutually personal, but not-objectifying relationship between an I and an Other, see Santmire, "I–Thou, I–It, and I–Ens."

21. I find the phenomenology of Teilhard de Chardin helpful at this point, illuminative of the biblical witness. Teilhard held that all creatures, even the most minuscule, have a "within," a certain kind of subjectivity (often it cannot be detected), as well as a

At the same time, as we will see presently, that general kind of communication also presupposes a kind of partnering with *all* creatures on God's part and even, on occasion, God depending on a variety of those creatures to respond by their own canons of spontaneity and praise.[22]

If the whole point of the story in Genesis 1 focuses on God, moreover, it becomes dramatically apparent right from the start that this is a God who indeed wants to have a history with a world of many creatures. So Genesis 1 does not begin as a dogmatic treatise might, with a "doctrine of God" (*locus de Deo*), but immediately shows us God bringing a cosmic history into being and becoming, and partnering with the many creatures who are thus called into existence.

Why all these creatures? Bertrand Russell once asked that very question: if the Bible is right, if humans are at the center of things, what are we to make of all the ichthyosauruses and the dinosaurs? Why did the Lord take such a long time to get to the main point of the project? It appears that Russell never really understood the Priestly witness. The Lord does indeed take a long time, as it were, to arrive at the human creation, but for a reason. The Lord, according to the biblical witness, is launching a history with the whole world, with many creatures, not just with the human creature. This is why we hear the ritualistic repetition of the phrase: "and God saw that it was good." Each stage, each day, of God's creative activity has its own integrity and its own meaning in the greater scheme of things.[23]

"without," an empirically identifiable kind of structure. That subjectivity becomes more and more definitive of creatures' identities the more complex they become empirically, in Teilhard's view, in particular as they develop central nervous systems and are "cephalized." In the human creature alone, Teilhard held, the "within" comprehends the "without," the within is *primary*—or it can and should be. Accordingly, the Divine personal communication with the human creature is, in terms I am using here, a *communion*, a personal *ad extra* relationship of God with the creature whose subjectivity is the whole which is greater than the sum of its parts—with the creature who is created according to the image of the personal God. This Teilhardian phenomenology meshes well, in my view, with Martin Buber's phenomenology of I and Thou. For an interpretation of Teilhard and Buber in more detail, see my two books, respectively, Santmire, *Travail of Nature*, ch. 8; and Santmire, *Nature Reborn*, ch. 5.

22. For further discussion of the theme of creaturely spontaneity, see Santmire, *Brother Earth*, 133–39. On the elusive, but important (especially for the theology of nature) biblical theme of nature praising God, see the seminal essay by Terence Fretheim, "Nature's Praise of God in the Psalms." Also see Santmire, "Two Voices of Nature."

23. Regarding the construct, the integrity of nature: as far as I know, Joseph Sittler and I were among the first, if not the first, theologians to make use of this construct theologically and/or exegetically, Sittler in a 1970 essay titled "Ecological Commitment

God chooses to engage Godself and to share the Divine life with all these creaturely domains, in their own right.

Yes, *humans* are created to "rule" over the earth (Gen 1:28)—more on this presently—but, likewise, in the same language, *the sun and the moon* are made to "rule" over the day and the night (Gen 1:16-18). "Rule" here means something quite different than some human will-to-power. Instead, we see here a vision of a beautiful, interrelated whole of many different creatures, all of which are created by God to have a history with God: which is, in so many words, the whole point of the whole project. When, to instance Bertrand Russell's kind of thinking again, God finally "gets around" to creating the human creature, be it noted that God does not rejoice over just the emergence of the *human* creature, as if that were the whole point of God's creativity (as some later Christian interpreters, like St. Ambrose, imagined): God saw "*everything* that he had made, and indeed it was very good" (Gen 1:31). The whole point of God's creativity is the flourishing—and, again, implicitly, the fulfilment—of the whole in all its diversity.

Likewise, in keeping with the motif of the goodness of every creature, God does not rush on through, as it were, the first five days. God does not instrumentalize or "thingify" what some might think of as the "lesser creatures," in order to enter into personal communion with the human creature, although, as we shall see, that special kind of relationship between God and the human creature is taken for granted by the Priestly writers. On the contrary, God respects all creatures, works with them, takes time with them, as it were, befitting their own created potential, in order to enhance and realize the integrity of the whole. God blesses *the fish and the birds* and calls them to participate in God's creative project: to multiply and fill the earth (Gen 1:20-22). *The waters*, even more dramatically, in view

---

as Theological Responsibility" (reprinted in Sittler, *Evocations of Grace*, 76-86), where he spoke of "honoring of the immaculate integrity of things which are not myself" (80). My discussion of the construct appeared in my programmatic study, Santmire, *Brother Earth*, ch. 7. Sittler and I then together made use of the term, as contributors, in the 1972 study guide, *The Human Crisis in Ecology*, ed. Franklin L. Jensen and Cedric W. Tilberg, published by the Board of Social Ministry, Lutheran Church in America, to accompany that Church's 1972 social teaching statement on the environment.

By now, the term "the integrity of nature" and related constructs have gained a wide currency in ecotheological circles. Perhaps the most succinct theological definition of the generic idea has been offered in McDaniel, "'Where is the Holy Spirit Anyway?' Response to a Skeptical Environmentalist," 165; for him, the term refers to "the value of all creatures in themselves, for one another, and for God, and their interconnectedness in a diverse whole that has unique value for God."

of the connotations of chaos they had in cultures of the time, collaborate with the Creator, by Divine invitation: "Let the waters bring forth swarms of living creatures . . ." (Gen 1:20). In a like manner, God calls upon *the earth* to "bring forth living creatures of every kind" (Gen 1:24). *Humans* are mandated by God, in the same way, to be participants in the Divine creativity by being fruitful and multiplying—and also, again, as we shall see, by ruling and subduing the earth (more on this text presently) (Gen 1:28). All creatures are, in this sense, some explicitly, others by implication, *partners* with God's creativity, not merely objects of God's creative will posited for the sake of God's relationship with humans.

Further, God is depicted as creating both humans and the animals, the wild and the domesticated, on the same day (Gen 1:24ff.), thus suggesting a certain kind of solidarity between the two kinds of creatures. This suggestion is underlined by the strong implication of the solidarity of non-violence: humans and animals are created to be at peace with each other and not to prey on each other (Gen 1:29–30). This is also a critical ingredient of the goodness of God's creative project that the Priestly writers envisioned. In that sense, God depends on the humans and all the other animals, right from the start, to establish God's creative purposes by eschewing violence.

On the other hand, humans alone are created according to the image of God, in the view of the Priestly writers; the other animals are not created according to that image (Gen 1:27). This surely suggests a special relationship between God and humans, which does not exist, as such, between God and the other animals. This is already signaled by the fact, noted by William Brown, that the creation of the humans is introduced as a unique product of Divine intervention: whereas the land-based creatures are products of the land (Gen 1:24), human beings are not. "The opening command," Brown observes, "is 'Let us make human beings in our image,' not 'Let the earth bring forth human beings.' Unlike the Yahwist's anthropogeny, the Priestly writer makes clear that the land is not the source of human identity but only humankind's natural habitat."[24] The God-human relationship is also understood to be reciprocally personal: here for the first time in the story of God's creative acts, God speaks in the first person (Gen 1:26). Here the Divine "I" calls the human "thou" not just into being and becoming in partnership with God, but into communion, into the intimacy of personal communication.

---

24. Brown, *Ethos of the Cosmos*, 43–44.

In this respect, the Priestly writers—especially *these* tradents, given their cultic interests—must surely have presupposed that the Divine-human relationship is one of self-conscious praise, on the part of humans. The thought of the coming Sabbath on the seventh day, as the appointed setting for the humans to glorify God for all God's good works, was undoubtedly not far from the Priestly writers' minds, as they shaped the construct of humans created according to the image of God. From this Priestly perspective, in other words, the relationship between God and the human creatures is teleological, in a way that God's relationship with the other animals is not: the Creator brings the human beings into existence, so that they may in some sense "image forth" God's purposes on the earth both by working to establish human community—by "making history," as Juergen Moltmann likes to say—and by self-consciously worshiping the Creator.

For sure, the Priestly tradents take it for granted that *God* alone is the Creator—as indicated by the oft-observed fact that the word for creating (*bara*) is used only for the creative activity of *God*, here in Genesis 1 and throughout the entire Old Testament. Clearly this project of cosmic creation is intended to be viewed as beginning and ending, and as sustained, by the creative power and wisdom and self-giving of the God of glory. It is radically theocentric in that sense. The Creator is the Creator, and the creation is the creation.

For this reason, the Creator is to be glorified, as the Psalmists often say, and as the Priestly writers surely believed, not the creation or any other supra-human powers that might be thought to contend with God.[25] This is one of the reasons why, later in the Old Testament story, worship of images is prohibited. In no sense is the creation itself Divine. On the other hand, by God's gracious engagement with, respect for, and, in the human instance, communion with, God's variegated creatures, they surely are intended to be viewed as having their own integrity and, in various ways, their own spontaneity and so their own goodness in God's eyes—and hence have their being and becoming as God's partners, each creature in its own way.[26]

So this radically theocentric project envisioned by the Priestly writers is also, in that sense, thoroughly cosmocentric and thoroughly anthropocentric. More precisely, it is profoundly *relational*—even ecological, if that

---

25. Cf. Brown, *Ethos of the Cosmos*, 42: "Given the rich ancient Near Eastern background behind the so-called *Chaoskampf*, the archetypal conflict between the Deity and chaos, the Priestly cosmologist boldly divests all intimations of conflict from divine creation."

26. Cf. the discussion of this point in Fretheim, *Suffering of God*, 73.

term from a quite different world of discourse be permitted here, exegetically—rather than exhibiting a kind of hierarchical, regal-command character. For the Priestly tradents, in this sense, God is profoundly *with* all His creatures, related to them and interacting with them as they respond to God's creative initiatives.

Terence Fretheim's summary of the Old Testament's view of God's creative presence with God's creation surely reflects, overall if not in every nuance, the witness of the Priestly writers in Genesis 1, in particular: "God is graciously present, in, with, and under all the particulars of his creation, with which God is in a relationship of reciprocity. The immanent and transcendent God of Israel is immersed in the space and time of this world; this God is available to all, is effective along with them at every occasion, and moves with them into an uncertain future. Such a perspective reveals a divine vulnerability, as God takes on all the risks that authentic relatedness entails."[27]

In this sense, the theology of the Priestly writers in Genesis 1 is *subversive*: it stands opposed, implicitly if not explicitly, to some of the most fundamental cultural imagery of the writers' own socio-political milieu. It has often been observed that the Priestly accounts of the creation were given their literary shape—although they contain materials from much earlier times and may have received their final editing much later—in the setting of the Exile, that is, in the context of Babylonian rule. And *that* society was a hierarchical, command society, without a doubt. For the Babylonians, the word of the monarch was law, absolutely, and that word dominated both people and nature at will, as it was implemented by the monarch's subordinates, who could readily be executed if they did otherwise.[28] Soberingly, historic Israel from the era of David and Solomon at least into the Exile, often took that kind of command royal ideology for granted and, with that, images of God as the chief monarch of the cosmos.[29] The relational, ecological vision of the Priestly tradents *contradicts* that ideology.

Walter Brueggemann has suggested instructively, in this connection, that this is due to an inner-theological dynamic. The vision of God

---

27. Fretheim, *Suffering of God*, 78.

28. Cf. Keith Whitelam, "Israelite Kingship," 121: "Royal ideology provided a justification for the control of power and strategic resources; it proclaimed that the king's right to rule was guaranteed by the deities of the state. A heavy emphasis was placed on the benefits of peace, security and wealth for the population of the state which flowed from the king's position in the cosmic scheme of things" (quoted in Habel, *Land Is Mine*, 17).

29. See Habel, *Land Is Mine*, ch. 2.

presupposed by the Priestly writers is very much like the vision of God presupposed by the prophet Ezekiel (chapter 34), who wrote in the same kind of socio-political context: for Ezekiel, God is the "shepherd King" who cares for God's flock.[30] God is not the absolute monarch, whose word dominates the whole realm. Further, an anti-monarchical polemic seems to emerge here in this Priestly setting, almost in so many words, and is taken for granted by the Priestly writers, in any case: insofar as *humans*, in particular, are said to be created according to the "image of God"(Gen 1:27). In the ancient Near-East, typically, only *kings* were thought of as bearing the image of a god or gods.[31] Thus, while the monarchical imagery in Genesis 1 is evident, even essential for the Priestly tradents in light of their faith in the power of the God of wisdom and mercy who creates by speaking, that imagery is indeed profoundly qualified by other theological assumptions, which keep this text well within the overall Old Testament and, indeed, the general canonical view of God as the God of self-giving love, a faith rooted in experience of the earliest of Israelite communities.[32]

It is in this exegetical context that the much-discussed theme of human dominion over the earth, announced by the Priestly writers in Genesis 1:28, should be heard. The words themselves tell a harsh story, as has often been noted. Dominion or "rule" (*rada*) generally means "exercise authority over" and "subdue" (*kabash*) literally means "tread upon." At the level of word study alone that would seem to imply—taken together with the idea that the human creature is to image-forth what could be thought of as the supposed absoluteness of a Divine monarchial rule—that God creates the

---

30. Brueggemann, *Genesis*, 32.

31. So Brown, *Ethos of the Cosmos*, 44. See also Brett, "Earthing the Human in Genesis 1–3," 77: "The characteristic association of the phrase 'image of God' with Mesopotamian kings and Egyptian pharaohs has been long observed, but the implications of this comparison have often been under-analyzed. If the health of the created order does not depend upon kings, then the democratizing tendency of Gen 1:27–28 can be seen as anti-monarchic. Indeed, there is an anti-monarchic tone to Genesis, which begins in Genesis 1 but extends into the second creation story and beyond. The polemical intent is subtle, but the evidence for it accumulates as the narrative unfolds."

32. Cf. Fretheim, *Suffering of God*, 128: "God is thus portrayed not as a king dealing with an issue at some distance, nor even as one who sends a subordinate to cope with the problem, nor as one who issues an edict designed to alleviate suffering. God sees the suffering from the inside; God does not look at it from the outside, as through a window. God is internally related to the suffering of the people. God enters fully into the hurtful situation and makes it his own. Yet, while God suffers with the people, God is not powerless to do anything about it; God moves in to deliver, working in and through leaders, even Pharaoh, and elements of the natural order."

human creature to dominate, even exploit the earth, as monarchs in the ancient Near East routinely did. But that kind of interpretation of Gen 1:28, as it stands, while sounding plausible, in fact appears to be more a matter of eisegesis than exegesis, once it is compared to much more plausible readings. Which brings us to the theme of these explorations here: not domination of nature, but creative intervention in nature—and much more.

To begin with, the socio-political setting of the Priestly writers in Babylon merits some attention. This was no simple agrarian society. An urban-centered agricultural society for the most part, it was both hierarchical and highly organized. This kind of society as a matter of course presupposed massive human interventions in the earth, above all through irrigation projects, in order to sustain its economy. In this socio-ecological setting, if there was going to be urban life of any scope, in contrast with simple agrarian life of small communities in regions like the hill country of Palestine, such large-scale interventions in nature were a *sine qua non* of social existence. One would expect economic realities such as those to be reflected in biblical texts that were shaped in such a socio-political world.[33] And indeed they are—and even, in one sense, are celebrated—as Brown explains, in contrast to the simple, agrarian assumptions of the Yahwist in Genesis 2 (to which we will turn presently): "Admittedly, the Priestly account acknowledges that human life in the land cannot exist in effortless harmony with creation; it can flourish only by establishing some measure of control over the earth. The Yahwist's notion of forcefully and painfully working the soil as a consequence of the curse is regarded by the Priestly narrator as a noble exercise."[34] Such human intervention in the earth is, as a matter of fact, for the Priestly writer *theologically* noble, since it represents carrying out of the particular partnership with God that is part of God's creative purposes: it makes the land "fillable" with human life, as Brown says—as anyone who has ever had any hands-on experience with the establishment of human community in some "untouched" natural setting will surely recognize, for example, laying foundations or drilling for water or clearing fields.

"Nevertheless," Brown observes pointedly, "such a commission does not require exploiting the earth's resources, as the specific language of

33. Although, cf. the caveat in Hiebert, "Re-Imaging Nature," 42: "In the preindustrial age of biblical Israel, it is impossible that the Priestly writer had more in mind in these concepts of dominion and subjection than the human domestication and use of animals and plants and the human struggle to make the soil serve its farmers."

34. Brown, *Ethos of the Cosmos*, 44.

subduing might suggest. The Priestly author gives clear contextual clues that clarify and qualify this dominion over the earth."[35] Brown suggests, for example, that the hoarding of resources by humans is implicitly forbidden, since the vegetation given by God for food is also given to the animals. More substantively, Brown explains: "As God is no divine warrior who slays the forces of chaos to construct a viable domain for life, so human beings are not ruthless tyrants, wreaking violence upon the land that is their home. By dint of command rather than brute force, the elements of creation are enlisted to fulfill the Deity's creative purposes."[36]

In order to underline this point, Brown instructively points to a later figure in the unfolding Priestly narrative, beyond Genesis 1—Noah. Brown observes that Noah "models primordial stewardship"—I would prefer to speak here in terms of *partnership*, a term that Brown apparently did not consider using in this connection—by sustaining

> all of life in its representative forms. His "subduing" of the earth entails bringing together the animals of the earth into his zoological reserve, a floating speck of land, as it were. By fulfilling humankind's role as royal steward over creation (1:28), Noah is a beacon of righteousness in an ocean of anarchy. Noah exercises human dominion over creation by preserving the integrity and diversity of life.[37]

Strikingly, a point not noted by Brown but very much in support of his claims here, Noah takes *both* the clean and the unclean animals with him on to the ark! Had his assignment been to "make this a better world," he surely might have seized upon this opportunity to leave the unclean behind—or the mosquitos, for that matter. But, on the contrary, Noah's vocation is to serve as a partner with God in behalf of the world that *God* created, with all its diversity, not first and foremost to improve the lot of humans on this earth.

Human engagement with nature is thus envisioned by the Priestly writers as within *limits*, both theocentric and cosmocentric. It could be called—*a limited partnership*. One could say, in this sense, that God expects humans, yes, to establish their own unique communities—to build—yet

---

35. Brown, *Ethos of the Cosmos*, 45.

36. Brown, *Ethos of the Cosmos*, 45. (Brown is using the word "command" here in a more positive sense than I have used it above, in my discussion of the authoritarian character of Babylonian rule.)

37. Brown, *Ethos of the Cosmos*, 60.

not with wanton destruction, but always in cooperation with and respect for all the other Divinely mandated domains of creation, each of which has its own intrinsic value, since it is valued itself by God.

This, then, is the consummately beautiful mosaic of God's creativity at the very beginning of all things, according to the Priestly writers.[38] This is why, all things, taken as a beautiful whole, each creature or creaturely domain with its own purpose in the greater scheme of things, all working together in majestic harmony, are seen by God, in the Priestly vision, as "very good" (Gen 1:31).[39]

That ordered, cosmic goodness is celebrated in many ways throughout the Bible, especially in the prayer book of ancient Israel, the Psalms. One Psalm, 104, is particularly worth recalling here, because, on the one hand, it appears to have had an evidentially close relationship to Genesis 1 and, on the other hand, because it sets forth the vision of God's beautifully diverse creation with lavish abandon, in contrast to the measured cadence of Genesis 1. Hence with this Psalm in view, our understanding of Genesis 1 will, to that degree, be both clarified and deepened, according to the traditional hermeneutical principle that "Scripture interprets Scripture."

The text of Genesis 1, in some form, and Psalm 104 may well have served originally as librettos for a festival in the Jerusalem Temple.[40] The Psalm may be read, in any case, as a kind of poetic commentary on the traditions that have been gathered in Genesis 1.[41] Here we encounter the grand vision of God's intimate involvement, God's partnering, with all creatures and God's presence to each appropriately, according to its kind.

---

38. Cf. Brueggemann, *Genesis*, 37, regarding creation as "good": "The 'good' used here does not prefer primarily to a moral quality, but to an aesthetic quality. It might better be translated 'lovely, pleasing, beautiful' (cf. Eccles. 3:11)."

39. Cf. Brown, *Ethos of the Cosmos*, 50–51: "A stable creative order prevails in this cosmos, accomplished not through conflict and combat but by coordination and enlistment. Each domain, along with its respective inhabitants, is the result of a productive collaboration between Creator and creation. The final product is a filled formfulness. Form is achieved through differentiation, the mark of goodness. While differentiating the various cosmic components, the process of separation, paradoxically, serves to hold the cosmic order together. Creations's 'filledness' is achieved by the production of life. From firmaments to land, boundaries maintain the integrity of each domain as well as provide the cement that binds the cosmos as a whole."

40. Anderson, "Introduction," 11, following Paul Humbert.

41. Levenson, *Creation and the Persistence of Evil*, 57, speaks of a "genetic connection between Genesis 1 and Psalm 104": "The correspondences are not total, of course, but they are impressive and cast heavy doubt upon the possibility of coincidence."

We see God wrapping Godself in light, as with a garment (v. 2), riding on the wings of the wind (v. 3), establishing the earth on its foundations (v. 5), speaking powerfully to rebuke God's thunder (v. 7), and making springs gush forth in the valleys (v. 10). We see the human community established by God in the midst of all this natural splendor and richness and beauty, blessed with a life of joy, with plenteous food and "wine to gladden the human heart"(v. 14). But that is not yet the end of the story, according to this poetic vision of God's creative activity.

Here emerges explicitly a theme that is implicit in Genesis 1 and which we will meet again, dramatically, in the poetry of the book of Job. The Psalmist takes it for granted that, given the magnificence and mystery of God's universal history with all creatures, there are times when humans' active engagement with nature will rightly cease and will rightly become one of—what I am calling in this chapter—awestruck contemplation.

This is the picture we see emerging here. God has purposes with all creatures that are often wondrous to behold in themselves (though on occasion even repulsive to humans), not just purposes that pertain to the human creature. God makes the high mountains for the wild goats (v. 18) and God makes the night, when humans have withdrawn, so that a whole variety of animals can come creeping out, when the young lions, in particular, "roar for their prey, seeking their food from God" (vv. 20–21). The note of violence here—lions seeking their prey—represents a view of primordial goodness that differs from the nonviolent vision of the Priestly tradents. We will meet this theme of nature red in tooth and claw in an even more vivid form in the narratives of Job. But it is important to note the contrast, already at this point, in two texts that otherwise have so much in common. The witness of the Psalmist and the Priestly writers stand in tension with each other at this point, in a way that may not even be complementary.[42]

Even more removed from the human world, and more wondrous and fearful to behold, according to the Psalmist, is the sea beyond, "great and wide," with "creeping things innumerable" (v. 25, NRSV). Strikingly,

---

42. It could be that the Psalmist and Job in a sense collapse the two-stage thinking of the Priestly writers, regarding this matter, into one stage. That is to say, for the Priestly writers humans are vegetarians before the fall and only permitted to eat meat after the covenant with Noah; the Priestly writers, in that sense, allow that violence in nature is, in that latter sense, Divinely ordained—and, presumably, also assume that, after Noah, violence among the animals is the Divinely permitted rule. For the Psalmist and Job—and for the Yahwist, according to Hiebert—the food chain, with some animals killing others, is given, right from the start, with the goodness of the creation.

God has God's own mysterious purposes with what for the Psalmist was the greatest and most awesome of creatures of the deep, the Leviathan, the gargantuan monster of the sea. God rejoices in this creature or plays with it (both translations are possible) (v. 26)! It is as if a poet in our time were to say that God rejoices in the billions of galaxies in our universe, indeed that God plays with them!

The Psalm then begins to conclude by celebrating, one more time, the immediacy of God's interaction with all of God's creatures, possibly with an allusion to Gen 1:1, where we see the "Spirit of God" (a possible translation, as well as "wind"), hovering creatively over the primeval waters: "When you send forth your Spirit they are created; and you renew the face of the ground" (v. 30).

After the Psalmist rejoices, one last time, in all this created glory and calls upon *God Godself* to rejoice in all God's works (v. 31), an ominous note is introduced at the very end, alluding to events in human history, the rampant human sinfulness that preceded God's decision to destroy God's own handwork with the flood, described after Genesis 1 by the Priestly writer: "Let sinners be consumed from the earth, and let the wicked be no more" (v. 35).

This is as far as the Psalmist takes us in his poetic commentary on the themes we know from Genesis 1. We are left with the vision of the great and wonderful and, in many ways, self-moving world of God's creation, in which God is immediately engaged, the whole of which is indeed very good, lavishly good in the Psalmist's view, yet not without its own kind of violence. In addition, a sobering hint of human malfeasance is introduced at the very end.

Genesis 1 itself goes further, at least explicitly, at this point: to a seventh day for God's creative project, a day which, although it stands in continuity with the others, is also quite different—the Sabbath. This is the day, we are told, when God rested from all God's creative activity (Gen 2:2). Here the accent shifts from goodness to *holiness*. "As all creation is directed toward completion," Brown explains, "completion sets the stage for consecration. Goodness and holiness, bounded and separate as they are, are also bound up in teleological correspondence, an integrity of temporal coherence. The primordial week, in turns out, is also a holy week."[43]

The meaning of the Sabbath, for the Priestly writers, is profound and complex, much too profound and complex to explore in brief compass

---

43. Brown, *Ethos of the Cosmos*, 52.

here. The theme of fulfilment is suggested, however, and that at least bears mention. While the whole creation, in the first six days, is *very good*, with the dawning of the Sabbath and the mystery of the Divine Rest itself drawing the whole creation to it, all things are in some sense to be sanctified, made *holy*, or perfected. That seems to be the Priestly vision. Interpretive thoughts such as these led Gerhard von Rad, in his Genesis commentary, to suggest that, whatever else the Sabbath might mean for the Priestly writers, there is a sense here—with a view to the unfolding history explored in Genesis 2 and beyond, above all with the awareness of the coming what has traditionally been called "the fall" and its aftermath—that the Sabbath as the eternal day of Divine Rest, and the perfection of all things in holiness, is a Day *yet to dawn fully* in this world. Hence the Sabbath can be interpreted as an *eschatological* Day.[44] Note that, in contrast to the other six days of creation, the Sabbath is never said to end. We do *not* read: "And there was evening and there was morning, the seventh day." In this sense, the Sabbath is ongoing. Even to eternity, to the last days? It would seem so.

So it is possible to hear this text about primordial beginnings suggesting also the promise of ultimate endings, pointing toward the time when perfect peace, *shalom*, will finally be established once and for all, when the universal history of God will one day be consummated, fully sanctified, beyond the sinfulness and the finitude of this world. In view of what we know is to come in the Great Story of God's universal history with the creation, then, as well as with respect to what we can hear from this Sabbath text itself, in its own historical setting, we can appropriately call to mind that kind of eschatological horizon here, if not read it directly from the text itself.[45]

---

44. Von Rad, *Genesis* 60–61: "that God has 'blessed,' 'sanctified' . . . , this rest, means that P does not consider it as something for God alone but as a concern of the world, almost as a third something that exists between God and the world. The way is being prepared, therefore, for an exalted, saving good . . . for the world and man . . . It is as tangibly 'existent' protologically as it is expected eschatologically in Hebrews (Heb., ch. 4)."

45. Regarding the emergence of an eschatological consciousness in the history of ancient Israel, in its own cultural setting, cf. the words of Schmid, "Creation, Righteousness, and Salvation," 110: "It has long been recognized that there is a close relation between views of creation and consummation [in the Old Testament]. The salvation (*Heil*) expected at the end of history corresponds to what the entire ancient Near East considered to be an orderly (*heil*) world, including the view of the pilgrimage to Zion to do homage to the God enthroned there as King . . . This is the new dimension in the eschatological horizon: in the course of time there was an increasingly sharpened awareness of the difference between the world of creation and that which can be realized in history. Consequently the period of salvation was postponed into an ever-receding future and eventually was expected to be the in-breaking of a completely new eon."

Later, for sure, in the visions of Isaianic prophecy, that explicit eschatological confession emerges: in the day of the promised "new heavens and new earth," all flesh will come to worship before the Lord, "from sabbath to sabbath" (Isa 66:22–23, NRSV).

Whether or not we understand the Sabbath in these eschatological terms, however, as we see it in its Priestly setting in Genesis 1, the second creation story in Genesis 2, to which we now turn, may be read as transpiring *on the sixth day*, as a fleshing out of the human story, in particular, in the midst of God's creation history narrated in Genesis 1. This thought, that Adam and Eve in the Garden and beyond, as the primordial characters in the human drama, lived on the sixth day and that indeed the subsequent unfolding of human history has occurred on the sixth day, was taken for granted by many early Christian interpreters of Genesis and assumed particular importance, later, in St. Augustine's theology of history.[46] Be that as it may, in reading the book of Genesis, we immediately come upon a second creation story in Genesis 2, which from the perspective of the final editors of Genesis could only have unfolded on the sixth day. This is the Yahwistic story of Eden and its aftermath.

The Witness of Genesis 2

This story complements the narrative of Genesis 1 in many ways, in some measure because the setting here is small-scale agrarian, rather than urban and institutional.[47] This is not to suggest that social settings necessarily determine theological meanings. It is rather to underline a commonplace of historical theology: that some theological affirmations sometimes emerge with much greater fluidity in some historical settings than in others. Such is the case here, with the evident agrarian setting of the story of Genesis 2.

To highlight the complementarity of Genesis 2 with Genesis 1, which is in many respects quite subtle, it seems advisable to step back from the Book of Genesis itself for a moment and to underline the terminology identified above, which is not given in Genesis, but whose meanings, nevertheless, in my view, leap out from these texts. Here we can begin explicitly to differentiate the first two dimensions of the Scriptural witness to God's intentions for humans' relationships with nature: first, *partnership with God and nature understood as humanity's creative intervention in the earth* and,

---

46. Santmire, *Travail of Nature*, 58.
47. See especially Hiebert, *Yahwist's Landscape*.

second, *partnership with God and nature understood as humanity's sensitive care for the earth.*

Genesis 1, we can say, projects a normative vision of the human relationship with nature in terms of intervention for the sake of justly and peacefully building human community: to fill the earth, in this sense, as the human family expands to all lands. Given the fact that the Priestly writers understood humans to be partners with God in carrying out God's purposes in this respect, we can gratefully recall the insightful terminology of the ecologist Renè Dubos when he describes what he calls "the Benedictine" approach to nature, to which he contrasts "the Franciscan" approach to nature, which in Dubos' view is more contemplative, predicated on respect and filial love.[48] While it would be anachronistic to project the image of St. Benedict and St. Francis back into the Old Testament, it is possible to invoke these constructs to help us with the interpretation of these biblical texts. The first, the Benedictine, refers to using nature appropriately, in partnership with God, for the sake of building human community all over the earth. The second refers to respecting and responding to nature, again in partnership with God, more in terms of nature's own needs. The two constructs are admittedly, in some respects, close in meaning,[49] yet they are also sufficiently different to be useful for biblical interpretation.

The Yahwistic creation story in Genesis 2, shaped by agrarian sensibilities as it is, definitely exemplifies what sensitive care for the earth can mean. As Theodore Hiebert has emphasized, for the Yahwist, "arable land is the primary datum in his theology of divine blessing and curse." In response to human sinfulness, the Divine curse diminishes the land's productivity, until the curse is lifted. God's blessing of Abraham is chiefly the gift of arable land. Also for the Yahwist, the three great harvest festivals of Israel shape the cultic calendar, and the primary cultic activity of these festivals is the presentation to God of the first fruits of the land and the flock.[50] So it comes as no surprise then to hear in the Yahwist's creation story that Adam is made from the earth—*adamah*. This is an observation that is frequently made, but Hiebert instructively wants to underline the concrete meaning

---

48. Dubos, *Reason Awake*, 126–27.

49. Cf. Brown's interpretation of what I am here calling "creative intervention" as it takes shape in the Priestly vision, *Ethos of the Cosmos*, 126: "God creates not by brute force but with great care. The human task of subduing the earth does not pit humanity against nature, but reflects a working *with* nature through cultivation and occupation, through promoting and harnessing creation's integrity" (italics added).

50. Hiebert, "Rethinking Traditional Approaches to Nature in the Bible," 28–29.

of that Hebrew word. Adam, it turns out, is not just created from the earth; he is created from the "arable soil." Such is the first human's agrarian identity, according to the Yahwist. "It is the claim that humanity's archetypal agricultural vocation is implanted within humans by the very stuff out of which they are made, the arable soil itself," Hiebert observes. "Humans, made from farmland, are destined to farm it in life and to return to it in death (Gen 3:19, 23)."[51]

For the Yahwist, it is almost as if God Himself were the premiere gardener! After forming the human creature from the arable soil, *Yahweh* "planted a garden in Eden," where Yahweh placed the human creature. Then "out of the ground the Lord God made to grow every tree that is pleasant to the sight and good for food" (Gen 2:7–9). Yahweh also, in due course, brings forth animals to be part of this landscape (Gen 2:19). The strong implication seems to be that Yahweh is involved in the care and the protection of this garden, setting the stage for the human creature to do likewise, as we shall see.[52]

Further, for the Yahwist *the land* is a character in its own right in this theological drama. The land has its own integrity, in this sense, its own essential place in the greater scheme of things. It is not just a platform to support human life. *The reason why the human is created, to begin with, is that there was no one to serve the land* (Gen 2:5)! So we see Yahweh forming the human from the arable soil—a theme that is missing from the Priestly account, as we saw, where the humans are created, as it were, directly—and then taking the human and placing him in the Garden of Eden in order to serve (*abad*) the land and protect (*samar*) it. The most familiar English translations of these words—"to till and to keep"—are profoundly misleading. The Hebrew tells a different story. The first term has the same Hebrew root as the word used by Isaiah to refer to "the *servant* of the Lord." The second term has the same Hebrew root as the word used in the Aaronic blessing: "May the Lord bless you and *keep* you." With only the received translation before them, general readers of this text might well understand it as a kind of capitalist manifesto: to develop the productivity of the land and keep the profits. They would have no reason to think that the words

---

51. Hiebert, "Rethinking Traditional Approaches to Nature in the Bible," 28.

52. Cf. Kahl, "Fratricide and Ecocide": "We might expect God to lean back and watch his creature taking up the spade to start digging and planting . . . But instead we see God taking the spade and planting the trees in the garden, definitely hard and dirty manual work . . . Adam's task is simply to serve and preserve the garden. Wherever humans touch the soil, God's footmarks and fingerprints are already there."

refer in fact to *identifying and responding to needs of the land itself and protecting the land from abuse or destruction.*

The image we have here is something like this: the experienced family farmer communing with the land — not too strong an expositional phrase in this context—down on his or her knees, gently transplanting a seedling, carefully finding a source of water for the plant, and then assessing ways to protect the plant from predators. Or we see the same farmer, thoughtfully and contemplatively watching over his or her olive trees, composting them or pruning them, whenever that seems appropriate. Here we see coming into view the sensitive (Franciscan) care of nature that the Yahwist champions, which both complements and stands in contrast to the Priestly writer's (Benedictine) vision of creative intervention in nature.

The Yahwist depicts the human's relationship to the animals, in much the same manner, in terms of tangible solidarity rather than intervention, certainly not any kind of domination. To begin with, both the human and the animals are made from the same arable soil (Gen 2:7, 19), a motif, as we have observed, that is missing from the Priestly narrative. Further, there is no apparent theological reason, as there was for the Priestly writers, sharply to define the differentiation between the two families of creatures, humans and animals, no "image of God" construct for the human in the Yahwist's view. Instead, the Yahwist is apparently quite comfortable with the thought that God makes both the human and the animal a "living soul" (*nephesh hayya*) (Gen 1:7, 19). One can recall here again that in traditional agricultural societies, humans and domesticated animals lived in very close proximity indeed, often occupying the same quarters. That kind of familial closeness is taken for granted by the Yahwist, as it also was, to some degree, by the Priestly writers, who envisioned the animals and the humans being created on the same day and who understood the humans to have been created as vegetarians.

The account of Adam naming the animals reflects the same Yahwistic assumptions, although the text has often been interpreted otherwise.[53] Many commentaries in the last two centuries routinely voiced the judgment, often drawing on examples from the history of religions, that naming is an act of power and that therefore Adam's naming of the animals was to be interpreted in terms of dominance.[54] The text, however, seen in its bibli-

---

53. See Ramsey, "Is Name-Giving an Act of Domination in Genesis 2:23 and Elsewhere?"

54. Brown, *Ethos of the Cosmos*, 141, seems to adhere to the older view, but this

cal context, actually tells a radically different story. In a certain sense, the Creator is depicted as withdrawing, for this moment: when Yahweh brings the animals to Adam to see what the human might name them (Gen 2:19). But this can be read as a thoughtful withdrawal to encourage creaturely bonding, on Yahweh's part, rather than, as it were, a disinterested deistic withdrawal whose purpose would be to hand over power to the human. The naming itself, moreover, can be understood as an act of affection on the part of the human, akin to the notion that Yahweh gives Israel, Yahweh's beloved, a name (e.g., Isa 56:5) or when Adam, rejoicing, gives the woman who is to be his strong, personal partner, a name (Gen 1:23). Comradeship on the part of Adam with the animals seems to be implied here in this naming scene, too, perhaps even with nuances of friendship and self-giving.[55]

All this—the human, formed from the arable soil, serving and protecting that soil and its lavish fecundities—illustrates why it is instructive to think of the Yahwist's vision of the Divinely given human relationship with nature as sensitive care for nature. There is even an implicit Priestly motif here: imaging God, an *imitatio Dei*. Did the Priestly tradents find this motif hidden here, when they edited the Yahwistic narratives, and then give it their own explicit articulation for their own reasons? Call it a kind of anticipation of 1 John 4:19, "We love, because He first loved us": we garden, because God first gardened for us. God plants the garden and then places the humans in it, as a blessing for the humans *and* as a calling, to serve and to protect the most fundamental stuff of the garden, the arable soil. To be faithful to that calling, humans will partner with God as God serves and protects the fruits of God's own creativity.

That the human is distinct from the animals, however, and destined for personal fellowship with God and other humans, in a way that animals are not, of that the Yahwist leaves us in little doubt. Adam finds no one with whom to commune among the animals. Adam only finds such a partner in the woman—exuberantly finds a partner in the woman—who is fashioned by Yahweh not from the arable land directly, as Adam and the animals were,

---

appears to be inconsistent with his overall approach to the theology of the Yahwist.

55. Note that Adam names only the living things, not all things. Cf. Westermann, *Genesis 1–11*, 229: "Names are given first to living things, because they are closest to humans." Cf. also the remarks in Fretheim, *Suffering of God*, 100, regarding the meaning of "naming" more generally in the Old Testament: "Giving the name opens up the possibility of, indeed admits a desire for, a certain intimacy in relationship. A relationship without a name inevitably means some distance. Naming the name is necessary for closeness."

but from Adam's own flesh. The idea of intense personal intimacy that is here suggested is sealed by the notion that the two are to be "one flesh" (Gen 2:24). The idea of the humans' intimacy with Yahweh is sealed, in a like manner, by the story of Yahweh conversing with them (e.g., Gen 2:16), as Yahweh does not do with the animals or any other creatures.

All this transpires in a setting of extraordinary natural fecundity, indeed in a garden of "delights," which is what "Eden" means. While Adam and then Eve are placed in that Garden to serve it and to protect it, there is no sense that that kind of daily work was to be burdensome for them—that kind of experience awaited them "after the fall." The Garden was a place of delights where they communed intimately with their Creator, who, as it were, walked with them in their life and work together in the garden, where they found bountiful and beautiful blessings in the creatures all around them, and where they lived at peace, in a certain kind of fellowship with all the animals. Although the Yahwist did not use these exact words to describe this primal scene, he very well could have depicted God at this point in his story, as the Priestly writers did in their own terms, seeing that all things were "very good."

Things did not turn out very well, however. What was intended by God, according to the Yahwist, went awry in the human domain. This, of course, is the story of what has been called traditionally "the Fall," recorded in Genesis 3.[56] This story is, as a matter of course, of critical importance for our explorations, first, because it is a grievous, but unavoidable chapter in what we are thinking of as the Great Story of the Bible, and second, more particularly, because it portrays destructive ramifications, as we will presently see, for the humans' relationships with nature, destructive ramifications which are only healed, finally, according to the Great Story, by the death, resurrection, and ascension of the Redeemer, Jesus Christ. If one of the results of the Fall of Adam and Eve is the Divine curse on the humans' relationship with nature, that all changes with the work of redemption, which is efficacious "far as the curse is found" (Isaac Watts).

Analysis of this *textus classicus*—the Fall—could take volumes, and, historically, has. I will limit the discussion to exploring the meaning of the Divine curse on nature. Specifically, what does it mean that, as they are excluded from the Garden by God because of their disobedience, the woman's pain in child-bearing is increased, as she also falls into a relationship of

---

56. For a nuanced exegetical interpretation of the meaning of "the fall" at this point, cf. Fretheim, "Is Genesis 3 a Fall Story?"

subservience to the man (Gen 2:16), and the "the ground is cursed" because of the man, and he is consigned to a life of painful toil (Gen 2:17–19)?

The woman's pain means at least this much: the pattern of domination of one person by another has emerged.[57] Likewise, the arable soil, once the congenial source of his life, now becomes a task-master for the man, a crushing burden. This is Brown's instructive summary of the meaning of the Divine curse for the Yahwist:

> The couple's disobedience has introduced not just the element of alienation, but also an ontology of bondage. Relationships between human beings and their environment are now based on power and control, as a matter of survival. As the man has been thrust into the harsh environment of the highlands of Canaan to eke out his existence, the woman is transported into the painful world of familial hierarchy and childbearing.[58]

The Divine curse is further intensified, according to the Yahwist, in the life of Cain, who killed his brother, Abel (Genesis 4). Now the "arable soil" itself takes on the role of juridical witness, according to the story, as it swallows up Abel's blood, and then demands redress. In response, God drives Cain from the "arable soil," the soil of his sustenance, into the vast and uncertain domains of "the earth."[59] This is his destiny, as Brown describes it: "Cain's exile is not from the human community per se. Driven from the ground, Cain is exiled to a social domain devoid of refuge and rife with violence, a realm of a social anarchy infinitely remote from the harmonious order of the garden."[60]

It is important to note here that there is no doctrine at this point, or elsewhere in the Bible, of any kind of "cosmic fall."[61] Sin *comes into* this primeval world by Adam's and Eve's "grab for wisdom," which was "an

---

57. Brown, *Ethos of the Cosmos*, 150: "In the world of curse, origin no longer indicates complementarity and mutual joy but domination and pain."

58. Brown, *Ethos of the Cosmos*, 150.

59. Brown, *Ethos of the Cosmos*, 168–69.

60. Brown, *Ethos of the Cosmos*, 169.

61. Cf. the comments of Terence Fretheim, *Suffering of God*, 39, citing the work of Rolf Knierim: "'History appears to have fallen out of the rhythm of cosmic order, whereas the cosmic order itself reflects the ongoing presence of creation. It remains loyal to its origin . . . And it knows about it.' Pss. 19:1–6, 103:19–22, and 148:1–6 are cited as examples of how 'the cosmic space proclaims daily and without end the glory of God, and itself as his handiwork.'" See, further, my reflections, "Biblical Thinking and the Idea of a Fallen Cosmos" [appendix], in Santmire, *Brother Earth*, 192–200.

outright betrayal of trust" in God.[62] Sin *results* in God's expulsion of the couple from their intended home of blessing, to a world of alienation from God and from each other and from the land, exemplified all the more dramatically by Cain's further expulsion into a world not just of alienation but of violence and chaos. The soil, in contrast, remains innocent, according to the imagination of the Yahwist. It protests against the violence of Cain. The soil remains the soil, even outside the Garden. It does not change. The Divine curse spills over on to it, because of the disobedience of humans and because of the fruits of violence that grow from that disobedience.[63] The Priestly writers take much the same approach to cosmic goodness and order: sin, for them, is clearly a social, not a cosmic reality.[64]

Which allows us to say, metaphorically, as we survey planet Earth today with the eyes of astronauts above, contemplating this beautiful, fragile blue and green island of life in the midst of the darkness of "outer space": we humans are living in Eden, yet behaving as if we were living outside of Eden. That the sinful violence of our lives, individually and collectively, sometimes pounds the earth and then rebounds back upon us with even greater destructive power—as in the case of climate change, for example, driven as it mainly is by consumer and corporate greed—is no fault of the earth.[65] The fault is all ours. And the rebound effect is a veritable Divine curse upon our sin.

Sadly, in our time, and perhaps it has always been so, that rebounding curse typically affects some more than others, above all the poor. Thus the impoverished masses of Bangladesh will in all likelihood be among the first to experience the mass devastations of climate change. That is why both the

---

62. Brown, *Ethos*, 157.

63. This is akin to the witness of some of the prophets, e.g. Isaiah 1:2 and Jeremiah 8:7, where "we find animals conforming to the will of God for their existence in ways not true of human beings" (Fretheim, "Nature's Praise of God in the Psalms," 29).

64. Cf. Brown, *Ethos of the Cosmos*, 129: "In the hands of the Priestly cosmologist, chaos is banished from the created order with the mere stroke of a stylus, put to rest, as it were. Rather than reifying, much less deifying, chaos as a necessary evil of cosmogony, Priestly tradition embeds chaos within the matrix of life itself, particularly human life, not as a necessity but as an ever present possibility. Chaos is violence run amok. It denotes the human violation of prescribed boundaries that foster the stability of community, a social contravention based on fear of and contempt for Yahweh's created order, in short, a desecration of creation and community."

65. Cf. the study of Fretheim, "Plagues as Ecological Signs of Historical Disaster," esp. 387: "the plagues . . . function in a way not unlike certain ecological events in contemporary society, portents of unmitigated historical disaster."

Priestly and the Yahwistic micro-narratives must be heard, especially by the prosperous, in conjunction with the strong voices of the prophets, so that the Priestly vision of a world full of justice and peace and the Yahwistic vision of a world where humans serve and protect nature will all the more powerfully claim, by the mediation of the prophets, our world of poverty and violence and looming ecological chaos.

The promise, according to the Great Story, is that, in Christ, with that deep human fault healed and the curse therefore existentially removed, we humans can begin to live in some measure as if in Eden again. Both the Priestly writers and the Yahwist give us glimpses of the kind of life that is possible in that very good world of God's own making, apart from the impact of human sin, which is the life that faith in Christ makes possible for us today, even as we continue to be immersed—willfully as well as by default—in a world dominated by the powers of sin, violence, and gross injustice.

The Witness of Job

The book of Job also gives us glimpses of human life in the very good world of God's creation. Yet with this micro-narrative, we encounter an almost entirely different way of seeing things. Call this a world at the edges of Eden. This is the lesser known side of the Job narrative. Best known is Job's own personal story of suffering and loss, a life of forced labor and no hope of liberation: "When I lie down I say, 'When shall I rise?' But the night is long, and I am full of tossing until dawn. My flesh is clothed with worms and dirt; my skin hardens, then breaks out again. My days are swifter than a weaver's shuttle, and come to their end without hope" (7:4–6). Thus stricken by outrageous fortune, Job angrily takes his case to God, and is berated by sages for doing so. But all this, anguished as it is, is but prelude for the place to which the Jobean narrator wishes to take us.

The narrator leads us into the experience of *wildness*, barely hinted at by the Priestly writers—although explicit, in some measure, in the creation theology of Psalm 104, that great poetic commentary on the vision of Genesis 1, when the poet talks about the lions roaring for their prey at night—and outside the imaginative purview, for the most part, of the Yahwist, as well. This is the world of nature beyond the creative intervention and the sensitive caring of human engagement and also, for that very reason, a world untouched by the Divine curse. This is the world of nature as God sees it and partners with it in God's own ways, apart from God's relationships with us humans and our lives in nature.

This is also a world where nature remains innocent, as it is for both the Priestly writers and the Yahwist. But this is an innocence that astounds, that overwhelms, and that even, at times, repels—especially when, seeing with the eyes of Job, we contemplate the pervasiveness of death in nature.[66] Here the themes creative intervention in nature and sensitive care for nature of the Priestly writers and the Yahwist give way to the theme of *awestruck contemplation* of nature. Partnership with God in the midst of nature and partnership with nature now mean stepping back from nature, *letting nature be and seeing it for what it is for God and in itself, apart from the interventions and the caring of humans.* This kind of partnership is, *mutatis mutandis,* akin to the partnership of a loving parent with an adult child, when the parent "lets go": when the parent steps back from the life of his or her adult child in times of challenge or trial, when the parent disengages, perhaps fearfully, but always with rapt attention.

The Book of Job, of course, is enormously complex and has profoundly puzzled many interpreters. Thankfully, however, a number of scholars have, in recent years, opened up the book in fresh ways, with sometimes compelling clarity. Brown is one of those interpreters, perhaps the most insightful. His approach is especially helpful for our purposes here, in any case, given his interests in the biblical theology of nature.

Although cosmology in the book of Job is all-encompassing, as Brown helps us to see, beginning as it does with the earth's foundation and the sea's fluidity, the voice that we hear speaking mainly addresses what might be called *the alien goodness of wildness.* This is Brown's summary of that Jobean vision:

> There, mostly wild animals, from lions to Leviathan, freely traverse the wasteland's expanse, sustained by Yahweh's gratuitous care and praise. The wilderness is where the wild things are, playing and feasting, giving birth and roaming, liberated from civilization

---

66. I take it that *death* in the sense of mortality, along with its anguish and pain, belongs to nature as created *good*. Note that nature is created very good, but *not* perfect, according to Genesis 1. That perfection must await the coming new heavens and new earth, when death will be no more. On the other hand, in the wake of human sin, death does become "the enemy" par excellence, in human history. It would take us too far afield, however, to argue this point here. For that kind of an argument see Santmire, *Brother Earth*, 124ff. and Terence Fretheim, "Is Genesis 3 a Fall Story?" 52. Also see Wilkenson, "Christian Theology of Death."

and ever defiant of culture, even in death. Undomesticated and unbounded, these denizens of the margins revel in their heedless vitality and wanton abandonment, unashamed and unrepentant of their unbounded freedom, which rests on a providence of grace.[67]

Experiencing a whirlwind of torment of his own, instructed unhelpfully by the counsel of sages, Job is driven into that world of wildness, and there he discovers who God is and who he is. The alien goodness of nature is, above all, expressed by the theophanous speech of God to Job, from the whirlwind. This speech, as Brown shows, has a twofold pedagogical purpose, both to broaden Job's moral horizon and to demonstrate Job's own innocence before the sages, who are his detractors. Although Brown does not use this language, it appears that there is something of Adam in Job,[68] before the fall, given Job's announced innocence. Thus God's speech itself never suggests any hint of punishment against Job. Be that as it may, Job encounters a world of innocence in nature, wild as it is.

In God's speech, God shows the care and precision with which the earth is established (38:4-7). "God is the architect," Brown points out, "and the earth is God's temple, not unlike the way in which the cosmos is patterned in Gen 1:2–2:3."[69] While the earth is thus a safe place, the sea is something else, in keeping with dominant apperceptions of the ancient Near East. The sea is depicted as flailing, like an angry infant, needing restraint. God, however, is up to the task. God fastens the doors to keep the sea from overwhelming the earth (38:10-11). Indeed, in keeping with the image of infancy, God appears as a midwife and caretaker of the sea, not unlike the role God assumed in Job's own birth (10:18). With the cosmos thus established, God leads Job into the wilderness.

This is indeed a wild place. In this Jobean discourse, we meet none of the images of cordial albeit fecund transformation of the wilderness that we do in prophets like II Isaiah, as Brown observes, the leveling of mountains or the raising up of the valleys (cf. Isa 40:4). This is nature as it is in itself, apart from human culture, raw and bloody, yet teeming with life, populated with exotic creatures appropriate to their respective domains. The animals appear two by two, lion and raven (38:39-41), mountain goat and deer (39:1-4), onager and auroch (39:5-12), ostrich and warhorse (39:13-25),

---

67. Brown, *Ethos of the Cosmos*, 394.

68. Janzen, "Creation and the Human Predicament in Job," 52, does make this connection.

69. Brown, *Ethos of the Cosmos*, 341.

and the hawk and vulture (39:26–30). Historically speaking, Brown notes, in the culture of the time "the animals highlighted in Yahweh's answer to Job were by and large viewed as inimical forces to be eliminated or controlled, an expression of cultural hegemony over nature within the symbolic worldview of the ancient Orient."[70] Kings, indeed, went forth to "conquer" such animals in ritualized royal hunts, in an effort to demonstrate by their triumph over these wild creatures their own ontic triumph over the forces of chaos, thereby establishing themselves, for all to see, as lords of both nature and culture.

Seen in this context, the Jobean discourse is radically counter-cultural. The great beasts of the wild, in this discourse, are indeed great and glorious and noble: images, Brown observes, that are "flagrantly at odds with their stereotypical portrayals attested elsewhere in ancient Near Eastern tradition."[71] The lion is an illustrative instance, the raven another. They are beautiful creatures in their own right, whom God feeds: "Can you hunt the prey for the lion, or satisfy the appetite of the young lions, when they crouch in their dens or lie in wait in their covert? Who provides for the raven its prey, when its young ones cry to God, and wander about for lack of food?" (38:39–41). Objects of contempt in the established culture of the time, the lion and the raven, Brown comments, are here transformed to objects of Divine compassion.[72]

The Jobean discourse goes still further in its celebration of the wild. Not only do we see noble, wild creatures, nurtured by God. We also see noble, wild beasts celebrated, precisely because they *resist* human domestication. The wild ox, for example, was profoundly feared. It used its horns for goring. Job is taunted by God, with a view to the otherness, the noble alienness of the wild ox: "Can you tie it in the furrow with ropes, or will it harrow the valleys after you?" (39:10). The ostrich is likewise paraded before Job: "It deals cruelly with its young, as if they were not its own; though its labor should be in vain, yet it has no fear . . . When it spreads its plumes aloft, it laughs at the horse and its rider" (39:16–18). This creature, Brown notes, "connotes joy unbounded; its wild flapping and penetrating laughter exhibit the throws of ecstasy, confounding Job's preconceived notions about the somber ostrich."[73]

---

70. Brown, *Ethos of the Cosmos*, 350.
71. Brown, *Ethos of the Cosmos*, 360.
72. Brown, *Ethos of the Cosmos*, 361.
73. Brown, *Ethos of the Cosmos*, 364. Cf. Janzen, "Creation and the Human

Celebrating Nature by Faith

In all this, Brown concludes, it is significant that the animals are not brought to Job, as they were to Adam in the Yahwistic creation story, for their naming. Rather, "he is catapulted into their domains. Instead of being presented with a parade of exotic animals, Job has come to see what they see, to prance with their hooves, to roam their expansive ranges, and to fly with the wings to scout out prey."[74]

Finally, the most alien creatures of all, the Behemoth and the Leviathan, emerge before Job, and are described in great and vivid detail. These are *God's* creatures par excellence, profoundly dwarfing Job, untouchable by any human reckoning. The Leviathan, in particular, is presented as king of kings, hugely proud and worthy of profound wonder and even fear.

Brown draws this insightful conclusion about the Jobean vision, in retrospect, in comparison with the culture of the time: "No longer are conquering and controlling nature part of the equation for discerning human dignity." Human dignity is precisely to be one of God's many creatures, never forsaken by God, appearances to the contrary notwithstanding. Job is able, therefore, in the end, to claim new meaning for his life before God (*coram Deo*), as one among many creatures, all of whom are God's children, all of whom have been nurtured and set free by God.[75] So, taking to heart the images of God's empowering of those animals thought to be lowly and God's nurturing of those animals thought to be mighty, Job returns to his own, Divinely created domain, the human community, but with a new self-understanding and a new awareness of the needs of others, especially the needs of those whom his society typically scorned or rejected. He begins his life afresh, as an alien in his own community. This is Brown's elegant picture of that return, here invoking, suggestively, the language of partnership:

> While Job does not forsake the wilderness, neither does he take up permanent residence there . . . Having become kin to these animals, Job retracts his patriarchy, both his honorable right to receive redress from the Lord of the whirlwind and his royal right to cultivate the nonarable landscape, and returns to his home and community, gratuitous of heart and humbled in spirit. Although

Predicament in Job," 51: "Over against . . . attempts to order and secure oneself and one's own in a dangerous world, the ostrich that lays its eggs on the unguarded ground constitutes Yahweh's description of birdly wisdom, a wisdom that appears as folly to the mentality of the Enuma Elish, the Baal myths, and indeed, aspects of the royal Jerusalem theology at least as popularly understood and acted on."

74. Brown, *Ethos of the Cosmos*, 365.

75. Brown, *Ethos of the Cosmos*, 375–76.

restored with a new family, Job is no longer willing to see the despised and the disparaged as objects of contempt. Like the animals, they are his siblings in the wild; they have become partners in a kinship of altruism.[76]

Having once been a stereotypical patriarch, then a social pariah, Job has now become, in Brown's words, "a vulnerable partner."[77] One can even think here of Job as a kind of "suffering servant" figure, paralleling or foreshadowing some of the proclamations of II Isaiah.[78] This is the legacy of the Jobean vision of awestruck contemplation of nature.

With reference to the Great Story as it can be told today, the Jobean vision can be read not only in terms of God's purposes with the wilderness areas of *this* planet—the fecund mountain ranges, the majestic oceans and their fragile coral reefs, the great whales and grand polar caps, the Siberian tigers, wildebeests, humming birds, and snail darters. It can also be read in terms of God's purposes with the "great things" of the whole cosmos, purposes that we can only barely begin to imagine—purposes with the billions and billions of galaxies, the super novas, the black holes, and the nearly infinite reaches of dark matter.

Even more, it can be read in terms of the final fulfilment of all things. Nature, too, as it groans in travail, has its divinely promised future, its final cosmic fullness and rest. Nature, including all the wild things and indeed all the galaxies and their mysterious cosmic milieu, writhes in anguished vitality, awaiting the day when, with human redemption finally completed, it will be able to reciprocate without bounds in its partnership with God, in the day when all things will be made new, when God will be all in all, when even the Leviathans of the cosmos will find perfect peace (cf. Romans 8:19–22).

---

76. Brown, *Ethos of the Cosmos*, 395.

77. Brown, *Ethos of the Cosmos*, 380.

78. So Janzen, "Creation and the Human Predicament in Job," 53: "The ancients were not wrong to conceive deity in royal terms. But they were wrong in supposing that royalty manifests itself in absolute invulnerability (or impassibility), and through overwhelming coercive power and aggressive control by means of a tight system, an airtight logic, of reward and punishment. The mystery of God's royalty is imaged in dust-and-ashes Job, suffering inexplicably, unshakably loyal to a God whom he does not as yet understand, and invited finally to share with God in the celebration and ordering of a world where the accepted risk of freedom is the creative ground of cosmic fellowship. It is not far from this to the astounding portrayal of Yahweh's 'arm' in Isaiah 53—the servant whose spoils of victory are won, not at the expense of enemy peoples, but on their behalf through unmerited suffering."

Celebrating Nature by Faith

Partnership With Nature in Retrospect

We have before us, then, this complex and rich biblical theology of partnership, between God and humans, between God and all creatures, between humans and every other creature, and between all the creatures of otherkind amongst themselves. That God has a partnership *with us humans*, and *we with one another*, is a thought that most students of the Bible in our time will take for granted. That God has a partnership *with nature* and humans *with nature* likewise and *all the creatures of nature are partners with each other*, are themes that may well need to be introduced to our Churches and to at least a few of our preachers and teachers. These theological thoughts, as it were, do not come naturally.[79]

But this is what the Bible shows us. God has a history with nature and values nature in itself, independent of God's relationship with the human creature. God creates a grand and beautiful world of nature for God's own purposes. It is harmonious. It is good. All its members are variously and sometimes exquisitely interrelated. But beyond its edges, it is also mysterious and even threatening to us. Which is to say, nature is finally God's business, God's infinite joy (cf. Ps 104:31), not ours.

God does fashion us and invite us, however, to be in partnership—a limited partnership—not only with God and with one another, but also with the beautiful and harmonious world of nature and to encounter its deep mysteries and its occasionally horrendous ambiguities. More particularly, God calls us to be in partnership with nature in three major ways, suggested by the Priestly writers, the Yahwist, and the narrator of Job.

There is indeed a time for everything, as we humans seek to partner with nature as well as with God and with each other—in this respect the witness of Ecclesiastes resonates with the Great Story of the Bible. There is a time to build, a time to care, and a time to contemplate. The witness of the Scriptures is that complex and that rich.[80] This complexity and this richness makes it all the more challenging for the faithful to take that witness to

---

79. There are historical reasons for this theological reluctance to focus on more than the Divine-human relationship. See Santmire, *Travail of Nature*, especially chs. 7 and 8.

80. In fact, it is much richer, as I indicated at the outset. A case in point is the passionate human sexuality celebrated in the poetry of the Song of Songs. This theme is something akin to what I have called awestruck contemplation of nature, but, as the biblical poet describes it, it is much more passionate and personal and intense, perhaps not like anything else in the whole creation. For this kind of reading of the Song of Songs, see Grossberg, "Nature, Humanity, and Love in Song of Songs."

heart, and for the Church to claim that witness in its own teaching and its own public testimony.

Thankfully, then, as Jesus Christ restores us to the place in God's history with the creation that God intended for us from the very beginning, Jesus also empowers us. He breathes the Spirit on us. And that Spirit will inspire us to discern which time is which: to read the signs of the times, as well as to hear the voice of the Scriptures in each of those times.

This is where the biblical theology of partnership leaves our Churches and our professional theologians and our preachers and our teachers and lay leaders and every faithful soul who loves the earth, finally, as we all seek to celebrate nature by faith in this era of global ecological emergency, rampant poverty, and mounting cosmic alienation.[81] Inspired by the Spirit, we will then be enabled to foster, whenever appropriate and in every complementary way, in partnership with God, with one another, and with nature as a whole: creative intervention in nature, sensitive care for nature, and awestruck contemplation of nature.[82] This is how we are rightly to live with nature, according to the Scriptures.

---

81. For a summary description of this global ecojustice crisis and cosmic alienation, see my book *Nature Reborn*, ch. 1.

82. This, of course, is a daunting challenge. But good, biblically conversant guidebooks are available to help us to discern the signs of the times and to find ways to appropriate and enact, both individually and corporately, the kind of exegetical findings that have emerged from this chapter. Among the best of those books are Bouma-Prediger, *For the Beauty of the Earth*; Hessel and Rasmussen, eds. *Earth Habitat*; Nash, *Loving Nature*. In this context, as in others, Pope Francis's encyclical, *Laudato Si'* is also essential reading.

2

Martin Luther's Theology of Nature
Announcing the God Who is in, with, and under All Things[1]

Martin Luther (1483–1546) celebrated nature by faith, profoundly and profusely. In various ways and at various times, he announced the God who is in, with, and under all things. For Luther, the world in which humans find themselves living, notwithstanding all its ambiguities and all its contradictions, is charged with Divinity and therefore, from his perspective, charged with promise. This may be the news report that our desacralized post-modern world-in-crisis needs to hear more than any other.

But interpreting Luther is an enormous challenge. To begin with, as I already have had occasion to observe, Luther has been claimed as the hero of a thousand causes. Luther has been envisioned as a destroyer of the Western Church, as a sex-driven pseudo-reformer, as the first "modern man," as a forerunner of Nazism, as an exemplar of the dynamics of twentieth-century ego-psychology, as a nature-mystic and visionary of the Ground of Being, not to mention as an audacious Church reformer, an insightful Bible commentator, and a theologian of the first order.[2] Which Luther are we to be concerned with in the following explorations?

---

1. This chapter is a thoroughly revised version of my previously published essay, Santmire, "Creation and Salvation according to Martin Luther," 173–202.

2. For a brief, but excellent description of Luther's historical identity, see Steinmetz, *Luther in Context*.

I suppose that if I were forced to answer that question, I would opt for Luther the audacious Church reformer, the insightful Bible commentator, the theologian of the first order. But that could sound like special pleading. Better, then to resolve to invoke the best canons of historical research and to let the findings speak for themselves.

A more subtle but still challenging problem for the kind of investigation that I am envisioning here is this. I am proposing to explore Luther's understanding of nature. But Luther did not use this term as we have become accustomed to it in the modern era of Western history. Hence the ordinary-speech understanding of the term that I am working with throughout this book will not quite work in this context. Nor will the biblical image "the earth," which was a suitable construct in the biblical interpretation in the preceding chapter. Luther uses the term earth only rarely, and then mainly in exegetical contexts. How, then, can we proceed with this investigation? I believe that we can, if we think first in terms that were familiar to Luther.

Those terms were creation and salvation, which, for Luther were comprehensive in meaning. In the overall scope of his thought, following biblical and traditional themes, from the Book of Genesis to the Book of Revelation, creation and salvation generally meant, for Luther, the coming into being, the history, and the consummation of all things, by the gracious workings of the triune God. I want to keep this universal picture in mind as we explore Luther's thought about what we moderns have thought of as nature.

Within the context of these larger understandings, it is then possible to think of nature as a subsidiary construct: as *the material-vital aspect of the triune God's creative and saving history with the whole creation*. This, I am aware, is a somewhat abstract way to talk about what biblical witnesses think of as "the earth" or what the theological tradition, following the Nicene Creed, has also has depicted as "all things visible." But I find it clarifying.

Introduction: An Elusive Theological Figure

Still, resolving to depend on the best canons of historical research, on the one hand, and defining nature in a way that allows us to grasp Luther's thinking in this respect, on the other, is only part of the challenge of interpreting his theology of nature. Luther is an elusive historical figure in many ways, theologically as well as otherwise.³ He did not produce a *Summa Theologica*, nor even an *Institutes of the Christian Religion*, as did Thomas Aquinas and John Calvin. Absent such a comprehensive work, it is therefore a challenge to identify Luther's theological baseline.

Calvin, for example, can be read as developing a comprehensive vision of a twofold knowledge of God *(duplex cognitio Dei)* over the course of many years, as he kept revising his *Institutes*.⁴ And that kind of vision of God as Creator and Redeemer gives a certain coherence to the entire argument of his *Institutes*. Luther seems to presuppose a similar epistemological understanding, but nowhere did he develop that thinking in any kind of extended treatise the way Calvin did.

Many of Luther's writings, moreover, were polemical and given to exaggeration. More than exaggeration, Luther on occasion could regrettably lapse into vitriol, as in his writing against the peasants and the Jews.⁵ Do such polemics belong to the core or at the periphery of his theology?

Further, a large portion of Luther's writings, as we have them, are commentaries or homilies on Scripture. Luther himself thought that, in those instances, he was discerning and announcing the Word of God, not sharing what he himself might have been thinking. Sometimes it takes considerable interpretive dexterity to tease out Luther's own meanings in such writings.

Perhaps the most substantive interpretive challenge before us, however, revolves around the primary constructs of Luther's theology, which I have already highlighted, creation and salvation. This is apparent when we consider the argument of a major modern study of Luther's thought, Bernhard Lohse's highly regarded investigation, *Martin Luther's Theology: Its Historical and Systematic Development*.⁶ Lohse maintains, in various ways, that Luther's theological genius, perhaps Luther's unique contribution to the history of theology, was to view everything theologically from the perspective of *salvation*.⁷ As Lohse's exposition unfolds, he accordingly gives

---

3. For a short introduction to Luther's life and thought, see Hendrix, "Luther," 39–56. The contemporary classic about Luther's life and thought is Oberman, *Luther: Man between God and the Devil*.

4. Dowey, *Knowledge of God in Calvin's Theology*. Dowey's thesis has been challenged in recent years.

5. Edwards, "Luther's Polemical Controversies," 192–208.

6. Lohse, *Martin Luther's Theology*.

7. In a chapter entitled "'The Uniqueness of Luther's Theology," Lohse, writes: "What is new [in Luther's theology, compared to earlier theological traditions] is that of all the questions with which theology must deal, the aim and goal in any given instance is the

very little direct attention (some 10 of 350 pages) to Luther's understanding of *creation*, and much of that brief discussion focuses on the human creature alone (anthropology).

This might lead the reader who is new to these matters to ask whether it is even possible to investigate Luther's understanding of the whole creation with any thoroughness. More particularly, regarding our explorations here, if Luther is not deeply concerned with the whole created order and focuses mainly on human salvation, where does that leave his understanding of what we think of as the natural world?

The answer is this: We need a more nuanced interpretation of Luther's thought about the whole creation than the one that Lohse has offered us so briefly. Luther's theology of human salvation is surely the foreground of his thought. This was the issue that preoccupied his reforming zeal and his theological struggles from his earliest years as a monk to the end of his life. But Luther—whose published works are enormous in quantity and diversity—could often wax eloquent, sometimes at great length, about the creatures of nature, especially in his biblical interpretation, works like his major commentaries on Genesis and the Gospel of John and also in his sacramental writings, a fact that a cursory reading of Lohse's book might not reveal.[8]

Hence while human salvation is surely the foreground of Luther's mature theology (and it was even more so the preoccupation of his earliest works), the whole creation is surely the background of that theology. And, as anyone who is aware of the history of art knows, on occasion the background of a painting is at least as important as, if not more telling than, the foreground. The background is not necessarily merely decorative. It can make its own statement, which brings vibrancy and richness to the whole. This is arguably one of those occasions in the history of theology.[9]

---

question of salvation. Questions about the doctrine of God, about the sacraments, about ecclesiology, can be dealt with only when this aspect is seen from the outset" (Lohse, *Martin Luther's Theology*, 35).

8. For general discussions of Luther's theology of creation, see Althaus, *The Theology of Martin Luther*, ch. 10; Kleckley, *Omnes Creaturae Sacramenta*; and Loefgren, *Die Theologie der Schoepfung bei Luther*.

9. For a brief treatment of Luther's thought and Calvin's, too, with particular reference to their thought about "nature" or the visible creation, set in the context of the history of Christian thought, see Santmire, *Travail of Nature*, ch. 7.

Celebrating Nature by Faith

The Underlying Dynamics of Luther's Thought: The Theology of the Word and the Sensibilities of Hearing and Seeing

Sometimes the shortest way there is the longest way around. Such is the case for this investigation of Luther's understanding of nature. We first must be able to grasp Luther's approach to the whole creation and its relation to—human—salvation, especially given the fact that those themes are related in his thought as foreground and background. That juxtaposition must be examined carefully and even-handedly, since it is a matter of some subtlety.

There were clear and pressing historical reasons why Luther placed so much emphasis on a proper understanding of human salvation and why creation tended to drift into the background of his thought. Luther, after all, fought a lifelong struggle against the Catholic Church of his time in favor of what he considered to be a proper understanding of salvation: by grace alone, through Christ alone, by faith alone, apart from works of the law. Luther's efforts to reform the Church were driven by that concern, and that was his constant preoccupation, both for polemical and pedagogical reasons, to the very end of his life. So it is not surprising that he was wont to state, in his own exaggerating fashion, that justification by faith is the chief content of theology.

And more. Luther anchored his reforming zeal in behalf of what he thought to be the right understanding of salvation by his powerful theology of the Word of God. He was, indeed, with a fervor and a constancy equaled by few other theologians in his time or before, a theologian of the Word.[10] This was the theological theme that undergirded his declarations about human salvation in the first place, but also, in a striking way, about the whole creation, as well.

Given with Luther's theology of the Word, furthermore, was the assumption that, for all intents and purposes, this is the only viable way for anyone to know God, that is, when God speaks.[11] On occasion, Luther could say that sinful humans, on their own, do have some knowledge of God, perchance as a creator or a lawgiver. But, in Luther's view, sinful humans, apart from God's Word, *de facto* live in a kind of inner terror, as they confront the nothingness of their lives in this world: their own existential *ex nihilo*. Even believers, in Luther's view, those who have heard and been

---

10. Pelikan, *Luther the Expositor*; Kleckley, *Omnes Creaturae*, 180–93.
11. Althaus, *Theology of Martin Luther*, ch. 6.

claimed by the Word of God, have no direct way to probe into the inner mysteries of God's life and being. God in Godself, for Luther, remains essentially incomprehensible. When all has been said, Luther's most powerfully voiced conviction therefore remains this: God is truly known only by God's revelatory Word.

This theology of the Word, in turn, was deeply rooted in Luther's theological sensibility, according to which he regularly championed *hearing* over against *seeing*.[12] That appears to be an obvious connection, for what is a spoken word, even the Divine Word, if it is not heard? And what sense would seeing make in a context that is defined by the speaking of the—invisible—God?

The sensibilities of hearing and seeing form a deep motif in Luther's theology and are fraught with many implications.[13] Hence it will be instructive to begin our investigation of Luther's understanding of creation in general and nature in particular by taking this long way around: by considering first the dynamics of his theology of the Word of God and, behind that, his particular theological sensibility which prompted him to prefer hearing over seeing, often, although not always.

We can see how Luther employs his theology of hearing strikingly in his commentary on the Fourth Gospel. John 1:14, about the Word being made flesh, is of course the *textus classicus* for the theology of the Word, and Luther does not miss any opportunity to celebrate the incarnation of the Word in Jesus Christ.[14] But, in commenting on the same Word, as the eternal Word of the Father, Luther writes just as forcefully in his commentary on John 1:1 about the power of the Word of God in the whole *visible* creation:

> God the Father initiated and executed the creation of all things through the Word; and he now continues to preserve His creation through the Word, and that forever and ever . . . Hence, as heaven, earth, sun, moon, stars, man, and all living things were created in the beginning through the Word, so they are wonderfully governed

---

12. In considering the sensibilities of hearing and seeing, we touch on a highly complex and much discussed subject in Western philosophy and theology. See Blumenberg, "Light as a Metaphor for Truth," 30–62; Chidester, *Word and Light*; Levin, ed., *Modernity and the Hegemony of Vision*; Miles, *Image as Insight*; Zachman, *Image and Word in the Theology of John Calvin*.

13. See, further, Graham, "Hearing and Seeing."

14. See especially Luther, *Luther's Works* [=LW], ed. Jaroslav Pelikan and Helmut Lehmann; LW XXII, 102–24.

and preserved through that Word . . . How long, do you suppose, would the sun, the moon, the entire firmament keep to the course maintained for so many thousands of years? Or how would the sun rise or set year after year at the same time in the same place if God, its Creator, did not continue to sustain it daily? If it were not for the divine power, it would be impossible for mankind to be fruitful and beget children; the beasts could not bring forth their young, each after its own kind, as they do every day; the earth would not be rejuvenated each year, producing a variety of fruit; the ocean would not supply fish . . . If God were to withdraw His hand, this building and everything in it would collapse . . . The sun would not long retain its position and shine in the heavens; no child would be born; no kernel, no blade of grass, nothing at all would grow on the earth or reproduce itself if God did not work forever and ever . . . Daily we can see the birth into this world of new human beings, young children who were nonexistent before; we behold new trees, new animals on the earth, new fish in the water, new birds in the air. And such creation and preservation will continue until the Last Day.[15]

I will come back to a consideration of Luther's construal of the Word of God in the visible creation presently. Here the point is to observe how Luther employs his theology of the Word in order to undergird not only his theology of human salvation, but also his theology of the whole creation. The theological background, creation, therefore emerges in Luther's commentary in striking fashion, no less than the theology of salvation, it would appear, because both themes are shaped by the dynamics of his powerful theology of the Word.

But what Luther gives with one hand he seems to take away with the other, precisely because of the sensibility that Luther presupposes when he develops his theology of the Word: the sensibility of hearing. Speaking-and-hearing, of course, is a characteristic, although not the only, mode of communication between persons. Non-human animals, however richly they may communicate, do not speak the way humans do.

Thus, for Luther, the Word of God addresses sinful human beings through Jesus Christ, the Word made flesh, a testimony that for Luther is mediated through the Scriptures; and those humans may then respond in faith and thereby receive the benefits of salvation. For Luther, the Word is known, moreover, only by the means of Grace, through the proclaimed

15. Luther, LW XXII, 26–27.

Word and the administered sacraments.[16] The Word of God thus is not known in the whole visible creation, as such, even though, as Luther thinks of these things, that Word is powerfully and creatively active there.

This is the underlying problematic of Luther's thought, then, with regard to the visible creation. Luther champions the sensibility of hearing, on the one hand; and often—although not always, as we shall see—he *rejects* the sensibility of seeing. Yet seeing is how persons primarily encounter the visible creation, as when Jesus says, "Behold the lilies of the field" (Matt 6:28; trans. Joseph Sittler). No wonder, then, that Luther's formative theology of the Word seems to be so precariously related to his theology of the visible creation. This will appear much more clearly, once we have considered some of Luther's own remarkable thoughts about the sensibilities of hearing and seeing.

For Luther, as many observers have noted, and as he himself often stated, "The ears are the only organ of the Christian."[17] This is a typical exegetical comment of the Reformer, regarding the account of Jesus riding into Jerusalem:

> But shut your eyes and open your ears and perceive not how [Christ] rides there so beggarly, but hearken to what is said and preached about this poor king. His wretchedness and poverty are manifest . . . But that he will take from us sin, strangle death, endow us with eternal holiness, eternal bliss, and eternal life, this cannot be seen. Wherefore you must hear and believe.[18]

Since for Luther, only the Word of God liberates us from sin and reveals to us the purpose of the Creator, "A right faith goes right on with its eyes closed; it clings to God's Word; it follows that Word; it believes that Word."[19] And that leads the believer into the darkness, where no one can see: "A Christian . . . is not guided by what he sees or feels; he follows what he does not see or feel. He remains with the testimony of Christ; he listens to Christ's words and follows him into the darkness."[20]

This emphasis on hearing was no less true of Luther's teaching about the sacraments, where one might have expected, if anywhere, to encounter

---

16. See Althaus, *Theology of Martin Luther*, ch. 6.
17. Luther, LW XIX, 1.1, cited in Miles, *Image as Insight*, 515.
18. Luther, WA XXXVII, 201–2, cited in Steinmetz, "Luther and Loyola," 5.
19. Martin Luther, *Werke: Kritische Gesammtausgabe* (=WA and WA-Tr [Tischreden]), WA XLVIII, 48.
20. Luther, LW XXII, 306.

a shift of sensibility from the moment of hearing to the moment of seeing. But Luther took up the traditional Augustinian view that the sacraments are "visible Words," but with the accent on the Word. "You should know," he stated characteristically, "that the Word of God is the chief thing in the sacrament."[21]

Luther carried through the same agenda architecturally, as well. On occasion, he could speak approvingly of images in the churches, but he generally did so only if those images had an evident pedagogical purpose, only if they communicated the Word.[22] Indeed, he liked to think of the Church building as a "mouth-house."[23] It would therefore not be an overstatement to characterize the earliest example of a surviving Church building designed and built within the Reformation milieu, dedicated by Luther in 1534, the small Schloss Kapelle in Torgau, Germany, precisely in those terms.

This was veritably a house for preaching. The elevated pulpit that dominates the plain space dramatically announces that the proclamation of the Word of God is that structure's primary purpose. The paintings that are in place, by Luther's friend, the well-known Lucas Cranach, depicted Luther preaching, among other images of worship in that space, such as the Lord's Supper. Those paintings were not first and foremost iconographical. They were primarily pedagogical. In that chapel space, therefore, the sensibility of hearing triumphed.

The contrast with Calvin, at this point, is illuminating. As a matter of fact, it could not be more dramatic. For Calvin, the sensibility of seeing is at least as important as the sensibility of hearing, if not more so. Thus Calvin can, like Luther, celebrate the Word of God incarnate addressing humans and also the eternal Word working powerfully throughout the whole visible creation. But Calvin affirms, as well, that the believer (and perhaps non-believers, too, to some extent) can contemplate—the right word—the glories of God throughout the whole visible creation.[24] (That Calvin also rejects visual images in a liturgical context, strikingly, is another matter, which need not concern us here.)

---

21. Luther, WA XLVII, 219.
22. See Koerner, *Reformation of the Image*.
23. Luther, WA X.1.1, 626, cited by Pelikan, *Luther the Expositor*, 63.
24. See especially Schreiner, *Theater of His Glory*. On Calvin and seeing, in particular, cf. Zachman, *Image and Word in the Theology of John Calvin*. For the unfolding of such themes in the works of Calvin's followers, see Lane, *Ravished by Beauty*.

Luther's accent on hearing, over against seeing, can also be instructively contrasted with the thought of Ignatius Loyola, as David Steinmetz did: "In my judgment, the most striking difference between the two theologians is the difference between Loyola's emphasis on the interior vision and the immediate experience of God and Luther's emphasis on the act of hearing and the mediation of the presence of God through a Word external to the self."[25]

But this is not yet to have penetrated to the heart of Luther's accent on the sensibility of hearing. For Luther, there are *dangers* embedded in the sensibility of *seeing*, profound dangers. This takes us to the foreground of Luther's reforming theology of salvation, to its major themes indeed, above all the theology of the Cross and the theology of glory, then to the theology of justification by faith alone.

Luther took the distinction between the theology of the Cross and the theology of glory for granted at least since the time of his 1518 Heidelberg Dissertation.[26] Luther understood the theology of glory to be a kind of speculative theology, which is predicated on theological interpretation of the created world. Had he wished to do so, Luther could have called this theology of glory, in his own terms, a theology of seeing. For this kind of theology, as Luther understood it, contemplates the created world and moves the mind, by that contemplation, to ascend ultimately to the vision of God and to an understanding of that God's purposes.

The theology of glory, in Luther's view, also leads inevitably, because of its rationalizing methodology, to a theology of justification by works, in one form or another. The theology of the Cross, in contrast, sees only the contradictory, inglorious vision of the Crucified, and knows God hidden and revealed there, by the proclamation of the Word of the Cross. That Word mediates the justifying Grace of God, according to Luther, which the believer receives by the hearing of faith.

From the perspective of the theology of the Cross, moreover, the theology of glory, which, in Luther's words, "sees the invisible things of God in works as perceived by man, is completely puffed up, blinded, and hardened."[27] Whether this core-distinction in Luther's thought allows any sustained contemplation of the larger visible creation, at this point in Luther's thought, seems *prima facie* doubtful. Contemplation of the visible

---

25. Steinmetz, "Luther and Loyola," 7.
26. Luther, LW XXXI, 40.
27. Luther, LW XIX, 52–53, cited in Lohse, *Theology of Martin Luther*, 38.

world of creation would appear to fall under the rejected rubric of the theology of glory.

The distinction between the theology of glory and the theology of the Cross is intimately related to another teaching of Luther, just alluded to, perhaps *the* core conviction of his thought, and so recognized by most students of Luther, a theme to which we have already referred to in a number of instances: justification by grace alone, through faith alone, apart from works of the law.[28] For Luther, "the article of justification is the master and prince, the lord, the ruler, and the judge over all kinds of doctrines; it preserves and governs all church doctrine . . . Without this article the world is utter death and darkness."[29]

To grasp the dynamics of Luther's thinking at this point, I want to invoke, once again, a term Karl Barth coined to describe his own thought, focused as it was on God and humanity, theoanthropology. The foreground of Luther's theology, as he himself describes it, is theoanthropocentric in this sense: it focuses on the justifying God and the justified sinner. Luther stated this clearly in a 1532 lecture on Psalm 51:

> Knowledge of God and man is divine wisdom, and in the real sense theological. It is such knowledge of God and man as is related to the justifying God and to sinful man, so that in the real sense the subject of theology is guilty and lost man and the justifying and redeeming God. What is inquired into apart from this question and subject is error and vanity in theology.[30]

With this core teaching of Luther in view, as when we considered his distinction between the theology of glory and the theology of the Cross, we may wonder what is to become of the larger visible creation in Luther's thought, since justification, as Luther understands it, seems to be almost exclusively theoanthropological, a matter of God and humanity alone.

All this appears to be the legacy of Luther's fateful celebration of the sensibility of hearing and his vigorous rejection—often, if not always—of the sensibility of seeing. The foreground of Luther's thought, his reflection about salvation, thus seems to stand in some real tension with the background, his reflection about the whole visible creation.

One might even conclude, in this respect, that the foreground of Luther's thought tends to overshadow the background completely in many

28. Lohse, *Theology of Martin Luther*, 74–78.
29. Luther, WA XXXIX.1, 205
30. Luther, WA II, 327–28, cited in Lohse, *Theology of Martin Luther*, 40.

instances, notwithstanding occasional expressions of interest by Luther, as in his commentary on John, in the creative Word of God in the whole visible creation. If we were to end these investigations at this juncture, indeed, we would have to conclude that Luther's theology offers us arguably a solid, rigorous, and liberating theology of human salvation, but that it has very little to offer those theologians today, who, in this our ecological era, are searching for a new and viable theology of the whole visible creation.

But that kind of conclusion would be premature and, in fact, *wrong*. For we have yet to examine the full sweep of the background of Luther's thought, his theology of the whole creation. When we do, we will encounter some remarkable theological claims on Luther's part. He develops, indeed, a visionary—the correct word—theology of creation, sometimes in what should probably be considered to be obvious settings, such as his magisterial commentary on Genesis, sometimes in what might appear to be improbable places, such as his sacramental writings.[31]

The Paradoxical Immanence of God in the Visible Creation

The theme of the Divine immanence has emerged in our time as critically important for those who are seeking to reclaim the biblical vision of the visible creation's goodness. Those who are concerned with issues in ecotheology have typically assumed—with some justification—that modern, if not all traditional, Christian thought has been focused mainly on the human creature, not on the whole visible creation also. With this conclusion: that modern Christian theology has been thoroughly anthropocentric.

The same theologians have also usually assumed—again, with some justification—that modern Christian thought, if not traditional Christian reflection, has focused mainly on the transcendence of the Divine, on the God who is "wholly other," in effect leaving the whole visible creation, if not the human creature in its midst, desacralized and disenchanted.[32]

Hence in our time there has been widespread theological interest, championed by theologians such as Sallie McFague and Juergen Moltmann, to reaffirm the goodness of the whole creation and the immanence of God in all things, so that the whole visible creation can be now be

---

31. See the extensive discussion in Kleckley, *Omnes Creaturae*.

32. For a review of these trends in theological modernity in the West, see Santmire, *Travail of Nature*, chs. 7, 8.

encountered—perhaps for the first time, they sometimes seem to suggest—as a world of Divine enchantment and Divine immediacy.[33]

This is one of the reasons why Luther's theology can, and perhaps should, appear so germane to contemporary theologians. Luther's theology is informed by a rich understanding of the Divine immanence and the fundamental goodness of the whole visible creation.[34] For Luther, in a sense, there is no other milieu known to us where God essentially dwells. We are not privileged to know about God's dwelling place in itself, in Luther's view. What we are privileged to know, rather, is that God is a present God, "God with us" and "God with all things," and therefore that the whole visible creation is profoundly good. Without compromising the Divine transcendence, as we shall see, for Luther God is wholly and immediately present, throughout the whole visible creation.

Luther developed his rich theology of Divine immanence in a variety of contexts, but with considerable intensity in his writings about the real presence of Christ in the Eucharist. Luther's theology of the Divine immanence has baffled some interpreters. But it is actually quite sophisticated, insofar as Luther invokes the discourse of *paradox*.[35] Paul Tillich, who considered himself to be writing in the Lutheran tradition, held that since theology has to do with both an infinite God and a finite world there is a logical place for paradox. Soren Kierkegaard, who also was giving voice to the Lutheran tradition in his own way, took the discourse of paradox with utmost seriousness, too.

In Luther's case, the language of paradox allows him to speak of God's presence both negatively—apophatically—suggesting, for example, that God is nowhere; *and* positively—kataphatically—saying sometimes, in the same breath, that God is everywhere.[36] Luther was at home in the tradi-

33. McFague, *Body of God*; Moltmann, *Trinity and the Kingdom*.

34. Cf. the statement in Steinmetz, "Scripture and the Lord's Supper in Luther's Theology," 266: "No theologian before or after Luther has celebrated the universal presence of God more than Luther has."

35. For a brief, but pointed discussion of the theological uses of paradox, see Walsh, "Paradox," 346–48. Also see the longer study: Hazelton, "Nature of Christian Paradox," 324–35.

36. Apophatic is from the Greek *apophasis*, referring to negation or "saying away." Kataphatic is from the Greek *kataphasis*, referring to affirmation or "saying with." The first has been used by scholarly interpreters to refer to spiritual or mystical experiences of "the ineffable," the second to spiritual or mystical experiences of what might be called "disclosures of the fullness of Being.'" See, further, Howells, "Apophatic Spirituality"; and Ruffing, "Kataphatic Spirituality."

tions of mystical theology, as a matter of fact, particularly as they were mediated to him through German theologians like Johannes Tauler.[37]

So Luther himself saw to it that an anonymous work in mystical theology, the *Theologia Germanica*, was published, and he praised it enthusiastically in an introduction.[38] He felt comfortable with mystical paradoxes, in particular, such as the saying of "the philosopher," Hermes Trismegistus: "God is the circle whose center is everywhere but whose circumference is nowhere."[39] And Luther was at home with the piety of the mystics, as when, early in his career, he spoke of the love for God as rooted in the inexpressible and the unfathomable God, which experience he called a "transport right into the midst of the innermost darkness."[40] For Luther, indeed, faith could be a kind of mystical rapture:

> The Christ faith is a being-taken-away (*raptus*) and a being carried away (*translatio*) from all that is experiential (*fuehlbar*) inwardly or outwardly to that which is experiential neither inwardly nor outwardly, toward God, the invisible, the totally exalted, inconceivable.[41]

Luther's most favored paradoxical practice in this respect was his invocation of *many prepositions* in order to point to the Divine immanence and to the Divine transcendence, as we shall now see.[42] Luther was by no means satisfied with the use of only a single preposition, such as "above" or "in," or even with the three prepositions which are most frequently associated in our time with his sacramental theology, "in, with, and under."

37. See Hoffmann, *Luther and the Mystics*.
38. Luther, *Theologica Germanica*.
39. Luther, WA–TR 2:1742.
40. Luther, LW XXV, 293–94, quoted by Hoffman, *Luther and the Mystics*, 85.
41. Luther, LW XXIX, 149, quoted by Hoffman, *Luther and the Mystics*, 163.
42. On occasion, Luther was wont to supplement his paradoxical discourse about the Divine immanence, using many prepositions, with abstract language that he inherited mainly from his training in the philosophy of nominalism. Thus he could distinguish between three kinds of Divine presence, circumscriptive, definitive, and repletive (for this, see Steinmetz, "Scripture and the Lord's Supper in Luther's Theology," 260). But it appears that the use of these terms was aimed more at his opponents in order to gain debating points, to add on still other reasons to support Luther's position. This debating technique was probably an afterglow of Luther's theological training in disputations (e.g. the Ninety-Five Theses). Luther's heart, however, seems to come alive when he discourses about the presence of God, using many prepositions. Cf. Steinmetz's comment (Steinmetz, "Scripture and the Lord's Supper in Luther's Theology," 261), that "the philosophical argument is not the most important one from Luther's point of view."

This places Luther's discourse about the Divine immanence at the opposite end of a metaphorical spectrum from a number of prominent twentieth and twenty-first century theologians, such as Sally McFague and Juergen Moltmann, who affirm what they think of as a *panentheistic* understanding of the Divine immanence. Their approach is actually a kind of *metaphorical reductionism*: the preposition "in" completely dominates their discourse. All things, it is affirmed, are "in" God, hence the term pan-*en*-theism. Not so for Luther. He insists that many prepositions *must* be used, this, in order to achieve what appears to be a highly suggestive integration of the apophatic and the kataphatic.[43]

As Luther does this, he radicalizes the idea of the Divine spatiality, too. Our commonsense spatial categories simply do not apply to God, he believes.[44] Thus the fullness of God can dwell in a single grain of wheat, yet be beyond all things at the same time:

> God is substantially present everywhere, in and through all creatures, in all their parts and places, so that the world is full of God and He fills all, but without His being encompassed and surrounded by it. He is at the same time outside and above all creatures. These are all exceedingly incomprehensible matters; yet they are articles of our faith and are attested clearly and mightily in Holy Scripture . . . For how can reason tolerate it that the Divine Majesty is so small that it can be substantially present in a grain, on a grain, through a grain, within and without, and that, although it is a single Majesty, it nevertheless is entirely in each grain separately, no matter how immeasurably numerous these grains may be? . . . And that the same Majesty is so large that neither this world nor a thousand worlds can encompass it and say: "Behold, there it is!" . . . His own divine essence can be in all creatures collectively and in each one individually more profoundly, more intimately, more present than the creature is in itself, yet it can be encompassed

---

43. Cf. Luther, LW XXXVII, 230, quoted in Bornkamm, *Luther's World of Thought*, 190: "Faith understands that in these matters 'in' is equivalent to 'above,' 'beyond,' 'beneath,' 'through and through,' and 'everywhere.'"

44. In this respect, as others, Luther was at home with the mystical tradition. Cf. the description of Meister Eckhart's discourse by Howells, "Apophatic Spirituality," 118: "Eckhart delights in wordplay designed to move his listeners and readers through affirmation to the negation of their understanding of God . . . , and then to a transforming 'negation of negation'. . . . This is especially evident in Eckhart's *German Sermons*, where metaphors are chosen for their ability to collide with and contradict one another, producing an 'explosion' of language as the meanings fall apart, out of which a new understanding is born."

nowhere and by no one. It encompasses all things and dwells in all, but not one thing encompasses it and dwells in it.[45]

This paradoxical use of multiple prepositions, with its radicalization of the idea of Divine spatiality, also distances Luther from a major Western metaphorical trajectory of a more traditional kind, evident especially in the thought of St. Augustine and his successors, the vision of the Great Chain of Being.[46] This vision presupposes, too, a kind of *metaphorical reductionism*. But in this case the single most important preposition is not "in," as in the case of those who argue for panentheism, but "above." God is envisioned as "the One," the purely spiritual being at the top of a hierarchy of being. And the way for the believer to encounter God is to keep "climbing the ladder," to keep ascending higher and higher, contemplating the multiple grades of being, from material things to living things, thence to the creature of embodied spirit, the human, and finally to and through the purely spiritual creatures, the angels, up to the highest, the One or "the Good" or God. The dominance of the metaphor of ascent is especially vivid in works like Dante's *Divine Comedy*.

True, this kind of conceptuality in Western theology typically presupposed that there is also a kind of Divine overflowing *down* the hierarchy of being, often thought of in biblical terms as the creation and conservation of the world, sometimes richly construed. Augustine, for example, can on occasion envision the immanence of that Divine overflowing in terms of the multiple use of prepositions, not unlike Luther's. But as Augustine was read by his successors, the end or *telos* of the soul always tended to be upward.[47] "Above" was the preposition that finally dominated all the others for many Augustinians. For Luther, in contrast, there is no "up"—just as there is no "in"—in any singular fashion. There is only up and down, in, with, and under, around and beyond, all referring to the believer's encounter with the visible creation, here and now, where the believer is standing.

Strikingly, Luther can speak of this paradoxical Divine immanence in the visible creation in tangibly *visual* terms. The sensibility of seeing seems to inform his discourse in such instances. Thus Luther can envision the whole creation visibly as the "mask of God."[48] This means, for Luther, that God is *hidden* there. But it also means that God is *powerfully present* there,

---

45. Luther, WA XXIII, 134–36, cited in Bornkamm, *Luther's World of Thought*, 189.
46. For this material, see Santmire, *Travail of Nature*, chs. 2, 3, 4.
47. Santmire, *Travail of Nature*, ch. 4.
48. Luther, WA XL:1, 94.

in front of your very eyes. In the same vein, Luther thinks of the Creator vividly as being "with all creatures, flowing and pouring into them, filling all things."[49] And God's activity never ends, according to Luther. God is "an energetic power, a continuous activity, that works and operates without ceasing. For God does not rest, but works without ceasing."[50] For Luther, therefore, creation is not merely some transcendental event before the beginning of time. The Divine act of creation is also *now*, as we have just seen and as we also saw earlier in Luther's commentary on John 1. And that Divine act intimately and powerfully permeates the whole visible creation.

Corresponding to Luther's understanding of the whole visible creation as permeated with the presence of God is Luther's view of the fitness of the visible creation itself to accommodate the Divine presence, a view sometimes identified by the theological formula that "the finite is capable of the infinite" (*finitum capax infinitum*).[51] This is an aspect of creation's goodness, in Luther's view. Luther was adamant about this theme particularly in his sacramental disputes, both against the Catholic side and against sacramental views such as those of Zwingli.

Luther believed that both the Catholic emphasis on transubstantiation (the replacement of the created substance of the bread and wine) and the Zwinglian emphasis on the Eucharist in terms of what it signifies (Christ cannot be really present in material elements and must therefore be in heaven), did not do justice to the goodness of the visible creation. As Kleckley observes, Luther held that both groups assumed that "it was undignified for Christ's body and blood to be attached to creaturely elements such as bread and wine," while Luther himself wanted to affirm that the elements of the visible creation are wholly congenial recipients of the divine presence and action.[52]

Also corresponding to Luther's view of the creation as permeated with the Divine presence is his consistently lavish praise of the visible creation as a world of wonder and enchantment.[53] If someone really understands a grain of wheat, Luther could say, that person would die of wonder.[54] While Luther does not generally present us with a view of the visible creation as

---

49. Luther WA X, 143.
50. Luther, LW XXI, 238, cited in Lohse, *Martin Luther's Theology*, 213.
51. See Hendel, "Finitum Capax Infiniti," 20–35.
52. Kleckley, *Omnes Creaturae*, 127. See, further, Kleckley's full discussion, 127–41.
53. For a discussion of these matters, see Santmire, *Travail of Nature*, 127–133.
54. Luther, LW XXXVII, 57–58.

"the theater of God's glory," as Calvin does, Luther, like Calvin, and indeed like numerous premodern theologians, has what can be called an "omni-miraculous view" of the created world. For Luther, the ordinary is in fact miraculous: "For the growth of the fruits of the field and the preservation [of them by God] of various kinds, this is as great as the multiplication of the loaves in the wilderness."[55] Thus, in one exuberant statement, which Kleckley chose as the title of his study, Luther could say that *all creatures are sacraments*.[56]

All this, the theme of *finitum capax infinitum* and the corresponding view of the visible creation as omni-miraculous, full of wonder and enchantment, is undergirded by Luther's theology of the creative Word of God, which we have already examined. For Luther, as Kleckley observes: "Because of the Word, creation is filled with wonders. When miracles occur, the power of the Word working in creation is the driving force. At the same time, the usual order that creation follows from its original plan is neither less miraculous because we have grown accustomed to it, nor is it less dependent on the Word for its continuing existence and maintenance."[57]

Throughout all his descriptions of the Divine immanence, it should probably be noted, in view of the criticism that has, on occasion, been directed at Luther in this respect, he guards the idea of Divine transcendence of the creation very carefully. Luther's use of many prepositions, and his regular observations that God cannot be contained by any creature or in all the creatures combined, guarantees that, for Luther, God is not to be thought of as in any sense being identical with the creation (*Deus sive natura*). God is not in the world as straw in a sack, Luther can say, in his own characteristic manner.[58] Luther is clearly *not* a pantheist, although some, especially those who are still thinking under the sway of the Great Chain of Being conceptuality, which envisions God's proper place as "above" all things, sometimes accuse Luther of that proclivity. Luther's apophatic use

---

55. Luther, WA XLIII, 139, 8-10; LW IV, 5.

56. Kleckley, *Omnes Creaturae*, 207.

57. Kleckley, *Omnes Creaturae*, 194.

58. Luther, WA XXVI, 339, 25ff., cited by Bornkamm, *Luther's World of Thought*, 188: "It is vulgar and stupid to suppose that God is a huge, fat being who fills the world similar to a sack of straw filled to the top and beyond . . . We do not say that God is such a distended, long, broad, thick, tall, deep being, but that he is a supernatural, inscrutable being able to be present entirely in every small kernel of grain and at the same time in all, above all, and outside all creatures."

of many prepositions decisively distances his thought from any pantheistic tendencies.[59]

The Believer's Duplex Relationship to the Visible Creation

How then does Luther think of the individual believer's relationship to the whole visible creation thus charged with the powerful, albeit paradoxical presence of God? Luther gives a variety of answers to this question, some of them negative, some of them positive. This reflects one of his best-known paradoxical teachings, that the believer is both sinful and righteous—*simul iustus et peccator*. Insofar as the believer is sinful, the believer's relationship to the visible creation will be *negative*, alienated. Insofar as the believer is righteous, the believer's relationship to the visible creation will be *positive*, in the process of being sanctified.

While Luther apparently never used the word duplex to refer to the believer's relationship to the visible creation, this expression captures his understanding of the matter very well.[60] It also reveals how suggestive Luther's thought can be, when read in terms of contemporary ecotheological concerns. Today many are searching for new understandings of what the human relationship with nature can be, in terms that are both realistic (*qua peccator*) and visionary (*qua iustus*).

First, recall Luther's understanding of the believer's negative relationship with the whole visible creation. For Luther there is, as it were, no direct

---

59. So Luther stated, LW XXXVII, 228, cited by Althaus, *The Theology of Martin Luther*, 107: "Nothing is so small but God is still smaller, nothing so large but God is still larger, nothing is so short but God is still shorter, nothing so long but God is still longer, nothing is so broad but God is still broader, nothing so narrow but God is still narrower. He is an inexpressible being, above and beyond all that can be described or imagined."

60. Although Luther was urgently concerned to announce what he considered to be a proper understanding of justification by grace alone, through faith alone, apart from works of the law, he also took it for granted that sanctification of the believer and good works would as a matter of course follow the sinner's justification. Thus he characteristically says of baptism, LW XXII, 287: "We are baptized in God's name, with God's Word, and with water. Thus our sin is forgiven, and we are saved from eternal death. The Holy Spirit is also bestowed on us; we receive a new nature, different from the one with which we were born." And LW XXII, 286: "[The Holy Spirit] also awakens in you holy and new thoughts and impulses, so that you begin to love God, refrain from all ungodly conduct, love your neighbor, and shun anger, hatred, and envy. Such works are performed by those who have been born anew, namely, born anew through baptism, in which the Holy Spirit is active, making new persons of them."

line of communion between the believer and the God who is in, with, and under the whole visible creation. That God is masked. Indeed, for Luther, that God—"the Majesty" as we have heard him say—is terrifying.[61] So Luther counsels the believer always to seek God in the Word of the Cross, not in the visible creation.

Further, the believer—along with all other humans—will as a matter of course encounter the whole world visible creation as *cursed* by God. Twenty-first-century ears typically are not be attuned to such language, but Luther takes it very seriously. While Luther's language has an alien ring for us, in this respect, it also seems to point to commonplaces in our experience of the natural world. In our own context, such talk about the Divine curse on nature might be translated, *mutatis mutandis*, into discourse about evolutionary violence, so-called "natural evil." Sometimes nature can be a curse to us humans, too, as in the case of tsunamis or plagues. So this kind of curse-language, notwithstanding its strangeness to modern ears, does seem to bring with it a certain descriptive authenticity.

Also, if humans are "curved in" upon themselves (*incurvatus in se*), as Luther was wont to suggest, we might want to think of the excesses of capitalism, let us say, which contribute to global warming, as calling down a curse, as it were, on all of us, especially the poor. In a word: "cursed be the planet because of us," especially those of us who hold the reins of power.

At the same time, we will finally be well advised, I believe, *not* directly to involve God in this equation the way Luther did. The latter kind of thinking typically leads to immoral excesses of its own, as when some preachers proclaim that the spread of the HIV virus around the world and the onset of AIDS everywhere is a sign of God's judgment on homosexual behaviors and also on illicit heterosexual acts.

Problematic for us as this kind of discourse is, for Luther the curse of God is straightforwardly the heritage of the fall of Adam and Eve. Especially in his Genesis Commentary, Luther can describe sinful humanity's relationship with the world of nature in great and vivid detail.

> . . . [T]he earth, which is innocent and committed no sin, is nevertheless compelled to endure a curse . . . [I]t will be freed from this on the Last Day, for which it is waiting . . . [Under the curse,] in

---

61. Cf. Luther, LW XXII, 308: "Therefore we should hold His [God's] voice, testimony, and statements in high esteem . . . God deals with us this way that we may be able to bear His presence. For if He were to come to deal with us in His true persona and majesty, we would be lost."

the first place, it does not bring forth the good things it would have produced if man had not fallen. In the second place, it produces many harmful plants, which it would not have produced, such as darnel, wild oats, weeds, nettles, thorns, thistles. Add to these the poisons, the injurious vermin, and whatever else there is of this kind. All these things were brought in through sin.

Luther believed, too, that the curse was made more severe following the time of what he reckoned to have been the Flood and, indeed, that the effects of the curse of God on the natural world were being felt with increasing severity in the days at the end of the world in which Luther believed himself to have been living: "The world is deteriorating from day to day."[62] So Luther concludes: "The closer the world is to its end, the worse human beings become. For this reason it also happens that harsher punishment are exacted from us."[63] Again, Luther's theological and cultural world differed profoundly from ours, but we may be better equipped to empathize with Luther's sensibilities, in this respect, than earlier generations, since we now appear to be living in an era when it is possible to imagine the death of nature.

Second now, consider Luther's understanding of the believer's positive relationship with the whole visible creation. In marked contrast with his ideas of the visible creation as the mask of—the hidden—God and the visible creation as cursed by God because of human sin, Luther can also imagine the believer—now as righteous and now as being sanctified, according to Luther's way of thinking—giving voice to paeans of praise for the goodness of God in the visible creation.[64] Luther's words in his explanation of the first article of the Apostles' Creed are typical of many other utterances, especially in his Genesis commentary:

> I believe that God has created me and all that exists; that he has given me and still sustains my body and soul, all my limbs and sense, my reason and all the faculties of my mind, together with goods and clothing, house and home, family and property; that he provides me daily and abundantly with all the necessities of life, protects me from all danger, and preserves me from all evil.

---

62. Luther, LW I, 206.
63. Luther, LW I, 216.
64. This is the finding of the thoughtfully argued essay by Gregersen, "Grace in Nature and History," 19–29.

> All this he does out of his pure, fatherly, and divine goodness and mercy, without any merit or worthiness on my part.[65]

In similar terms, Luther sees the creative Word of God providing the grain that sustains human life:

> We are ... to praise and thank God for making the grain to grow. We are to recognize that it is not our work but His blessing and gift that grain, wine, and all kinds of crops grow, for us to eat and drink and use for our needs, as is shown in the Lord's Prayer when we say: "Give us this day our daily bread." ... [W]hen we see a whole field or one grain, we should recognize not only God's goodness but also His power ... And with [that] great power He protects you [Luther says to the wheat itself]! What dangers have you not survived from the hour when you were sown until you are put on the table![66]

In this sense, for Luther, the Grace of God is experienced in creation as well as in salvation, for those who have the eyes of faith to see.

Luther's understanding of the traditional theological view of human dominion over the visible creation is set in the context of this rich vision of the Divine immanence and the believer's positive experience of the visible creation as a vital and variegated gift of Divine Grace. Notably, in such settings, Luther is thinking of the overflowing goodness of God in the visible creation, rather than assuming that the visible creation is some kind of depository that is given to human creatures for their own purposes. True, Luther does restate a traditional theological view of human dominion over the earth, dependent on the words of Genesis 1:28–30: "you must use the things given and granted to you by God in His kindness. You must rule, work, and strive not to tempt God."[67] But this kind of statement is never translated by Luther into any kind of mandate for the economic transformation of nature.

As a citizen of the sixteenth century, moreover, Luther obviously did not have a twenty-first century ecological consciousness nor, with that, was he aware of the radical transformations and, indeed, desecrations of the natural world that would result from the triumph of the spirit of modern capitalism. Nevertheless, Luther was highly critical of many of the policies of early capitalism, particularly that system's exploitation of the poor.

---

65. Luther, "Small Catechism," in Tappert, ed., *Book of Concord*, 345
66. Luther, LW XIV, 122.
67. Luther, LW V, 256.

Luther's best ethical counsel, indeed, was that believers should always focus on serving the neighbor, which was an ethical stance that he distinguished from the spirit of profit and the drive to acquire goods.[68] Luther understood human "dominion," then, *primarily in terms of service to the neighbor in need*, as he wrote in his Genesis commentary: "God does not give out his gifts so that we can rule and have power over others or so that we should spurn their opinion and judgment: rather so that we should serve those who are in such a case as to need our counsel and help."[69]

Such is Luther's vision of the overflowing presence of God in the visible creation and the blessed life of those who have been made righteous and who therefore can see that they are immersed in the goodness of that overflowing.

And while Luther never fully developed an explicit "cosmic Christology"(more on this presently) to complete his vision of the visible creation and the life of the believer therein, he did give us hints that such a Christology would be concordant with his rich vision of creation. The believer, who now is "in Christ," Luther believed, is given a new and more vital, even a contemplative relationship with the whole visible creation. Thus Luther writes in his commentary on John:

> Now if I believe in God's Son and bear in mind that He became man, all creatures will appear a hundred times more beautiful to me than before. Then I will properly appreciate the sun, the moon, the stars, trees, apples, pears, as I reflect that he is Lord over and the center of all things.[70]

By 1544, in a sermon, Luther can even somewhat self-consciously shift his discourse from the sensibility of hearing to the sensibility of seeing, alluding as he does to the traditional theological image of the two books of God, the book of Scripture and the book of nature:

> Our home, farm, field, garden, and everything, is full of Bible, where God through his wondrous works not only preaches, but also knocks on our eyes, touches our senses, and somehow enlightens our hearts.[71]

---

68. Lindberg, "Luther's Struggle with Social-Ethical Issues," 165–78.
69. Luther, LW II, 239
70. Luther, LW XXII, 496.
71. Luther, WA XLIX, 434, cited in Gregersen, "Grace in Nature and History," 28.

In the very last year of his life, remarkably, Luther self-consciously invoked the sensibility of seeing, as he wrote a notation in a volume of Pliny's works: "All creation is the most beautiful book or Bible; in it God has described and portrayed Himself."[72] In this sense, for Luther, the justified sinner who has heard the Word of the Cross can begin to see the whole visible creation with new eyes. *The sensibility of hearing has now been dramatically complemented by the sensibility of seeing.*

Further, from the perspective of the believer, who is in Christ, Luther also seems to have found a way to see the world as God originally intended it to be, "before the fall." Luther at times can present an almost Franciscan vision of that primal human identity in the visible creation, wherein the sensibility of seeing is, as it were, natural:

> If only Adam had not sinned, men would have recognized God in all creatures, would have loved and praised Him so that even in the smallest blossom they would have seen and pondered His power, grace and wisdom. But who can fathom how from the barren earth God creates so many kinds of flowers of such lovely colors and such sweet scent, as no painter or alchemist could make? Yet God can bring forth from the earth green, yellow, red, blue, brown, and every kind of color. All these things would have turned the mind of Adam and his kin to honor God and laud and praise Him and to enjoy his creatures with gratitude.[73]

Luther especially relishes the beauties of human corporeality before the Fall. So in his Genesis commentary, where countless doctors of the Church had sung the praises of human rationality, under the rubric of the image of God, Luther states that the fact that Adam and Eve walked about naked was their greatest adornment before God and all creatures. In the same vein, Luther envisions Adam and Eve as enjoying a "common table" with the animals before the fall.[74]

Finally, Luther further envisions the sanctified life of the believer, who has been justified by faith, as standing on the threshold of the eschatologically consummated creation. In the following utterance, the sensibility of seeing has definitely taken over Luther's thinking. And his eschatological consciousness has here been shaped by hope and joy, as distinct from the moods of fear and foreboding which sometimes seem to inform Luther's

---

72. Luther, WA XLVIII, 201.5, cited in Bornkamm, *Luther's World of Thought*, 179.
73. Luther, WA-TR IV, 198, cited in Steinmetz, *Luther in Context*, 24–25.
74. Luther, LW I, 42.

mind and heart most deeply, especially when he is agonizing about the curse of God on nature:

> We are now living in the dawn of the future life; for we are beginning to regain a knowledge of the creation, a knowledge forfeited by the fall of Adam. Now we have a correct view of the creatures, more so, I suppose, than they have in the papacy. Erasmus does not concern himself with this; it interests him little how the fetus is made in the womb . . . But by God's mercy we can begin to recognize his wonderful works and wonders also in flowers when we ponder his might and goodness. Therefore we laud, magnify, and thank him.[75]

All this Luther sees—again, the proper word—coming together and majestically transfigured on the Day of the consummation of all things, beyond every hint of Divine judgment:

> Then there will also be a new heaven and a new earth, the light of the moon will be as the light of the sun, and the light of the sun will be sevenfold . . . That will be a broad and beautiful heaven and a joyful earth, much more beautiful and joyful than Paradise was.[76]

This vision of the end time articulates and completes Luther's occasionally lavish statements about the sanctified life of the justified sinner in the world of the visible creation.

The Relationship between the Whole Creation and Human Salvation: The Narrative of Faith and the Grace of the Triune God

How, finally, does Luther generally relate the whole creation and human salvation, background and foreground? This is a critically important question for anyone who is seeking the grasp Luther's thinking about nature. Is nature, qua creation, merely some kind of stage and scenery for human life and salvation? No, there is no sign, even any nascent expression, of that idea so treasured by modern Protestant theology (Brunner, Barth[77]). Luther thinks much more simply in this regard—but not necessary less

---

75. Luther, WA-TR I, 1160, cited in Bornkamm, *Luther's World of Thought*, 184.
76. Luther, LW XII, 119, 121.
77. Santmire, *Travail of Nature*, chs. 7, 8.

compellingly—in narrative terms. His own self-understanding, after all, was that he was a Scripture scholar. And, as a matter of course, he took the grand narrative of biblical faith from Alpha to Omega—creation, reconciliation, redemption—for granted as the last word of biblical faith.

So Luther's theology was not systematic in this sense: it was not "faith seeking understanding" (*fides quaerens intellectum*) as it was for Anselm or, later, for Barth. It was "faith telling the story of the mighty and gracious acts of God in creation and salvation." Whether his theology was the stronger or the weaker for not seeking systematic understandings and for resting, instead, in the flow of the biblical narrative cannot be argued here. That, however, Luther's theology had its own strengths, in its own narrative terms, should be clear by now.

That Luther did not always give voice to that grand narrative in its fullness, moreover, should also be clear. There were times, as we saw, when the theological foreground of his thought, the theology of salvation, virtually obscured the theological background, the theology of creation. There were times when the sensibility of hearing, which was so critical for Luther in shaping his understanding of salvation, virtually replaced the sensibility of seeing, especially when Luther was rejecting the theology of glory in favor of the theology of the Cross.

In this sense, Luther's theology was indeed much more "existential" than "sapiential," to use the terminology of Otto Pesch.[78] Luther said, memorably, of the theologian: "It is not by reading, writing, or speculating that one becomes a theologian. No, rather, it is living, dying, and being damned that makes one a theologian."[79] Luther never thought of himself as projecting a grand and coherent system of theological truth as Thomas or even Calvin did, in their own ways; certainly not as Barth has done in our era.

But Luther was a theological visionary also, in the strict meaning of that expression. For all his passionate statements about hearing being the royal road to faith, and seeing being a dead-end for the emergence of true faith, Luther nevertheless envisioned—the proper word—a grand unfolding of creation and salvation history, in the midst of which the believer, now being sanctified, could in some measure see the grandeur and the beauty, as well as the power of God in, with, and under and also above and beyond, throughout and around, encompassing and unencompassed by, the whole visible creation.

78. Pesch, "Existential and Sapiential Theology."
79. Luther, WA V, 163.

And still more. Luther held creation and salvation together not only narratively, by presupposing a grand vision of creation and salvation history, modeled on the unfolding of the biblical narrative, from Genesis to Revelation. Luther's understanding of the relationship of the theological foreground, human salvation, and the theological background, the whole creation and its history with God, was also deeply and firmly predicated on what was an integrating thematic of his theology, the Grace of God or the self-giving love of God.[80] This thematic also shows us how integral Luther's theology of creation was to his thought as a whole, notwithstanding creation's evident background status.[81]

To grasp this thematic, we can instructively return to consider a motif from Luther's thought that we have already noted, the duplex relationship of the believer to the creation, insofar as the believer is both sinful and redeemed. The God whom sinners encounter, according to Luther, is masked, wrathful, and distant. The God who is revealed to believers is gracious, life-giving, and immediately accessible. "God is truly known," says Luther, "not when we are aware of his power or wisdom which are terrible, but only when we know his goodness and his love."[82] And this knowing comes only through Christ: "There is no other God apart from this Christ who has become our light and sun . . . He and no one else is the true God. It is he, I say, who has enlightened us through his Gospel."[83]

---

80. This kind of descriptive statement of Luther's thought is legitimate and instructive, even though it has a diagnostic character that does not do full justice to Luther the passionate, existential theologian. If Luther would have been asked about the central theme of his theology, he would in all likelihood answered "the theology of the Cross." For this kind of reading of Luther, see Pelikan, *Luther the Expositor*, 155.

81. Luther's understanding of the unity of creation and salvation in terms of the Grace of God was more subliminal, it appears, than systematic, more a matter of presupposing than explicating. It by no means had the clarity of one modern theological voice, that of Karl Rahner, but it circulated in the same orbit as Rahner's view: "We can understand creation and incarnation as two moments and two phases of the one process of God's self-giving and self-expression, although it is an intrinsically differentiated process" (Rahner, *Foundations of Christian Faith*); cited by Edwards, *Jesus the Wisdom of God*, 35.

82. Luther, WA II, 141, quoted in Althaus, *Theology of Martin Luther*, 191.

83. Luther, WA XXXI, 1, 63, cited in Althaus, *Theology of Martin Luther*, 191. Cf. also WA 19, 492; 19-24: "Although God is everywhere in created beings and I would love to find him in stones, or in the fire or in the water where he indeed is, but he does not want me to look for him there . . . He is everywhere, but he does not want that you are looking for him everywhere, only where the word is, there you must look for him and you will comprehend him."

This thought then takes us, with Luther, into one of the most subtle dimensions of his thought, which holds salvation and creation together intimately and inseparably, Luther's *Trinitarianism*. It is as if we were contemplating a Van Gogh painting. In the foreground we see two peasants standing in the shade of a haystack apparently at midday, the background being another haystack in that bright hayfield, all framed by an electric, blue sky. The whole scene is bathed in the brilliant rays of the sun. But the effervescent colors, which give the entire painting its vibrancy, are all the same, in the background as well as the foreground. Informed by his vision of the universal Grace of God, Luther's Trinitarian thoughts are the bright colors of his theology, which unite foreground with background.

In salvifically experiencing the Son, Luther holds, believers are led to the Father. "Scripture begins very gently," Luther explains, "by leading us first to Christ as to a man and afterwards to the Lord of all creation and finally to a God. Thus I come in easily and learn to know God. Philosophy and the wise men of this world, however, want to begin at the top and have become fools in the process. One must begin at the bottom and afterward rise up."[84]

Having begun with Christ, furthermore, the believer discovers that, as it were, *all is Grace*, not just salvation, but creation as well.[85] For Christ discloses the "fatherly heart" of God. And the compassion of that fatherly heart is given expression *everywhere*. So Luther comments on the story of Jesus' baptism, wherein the voice of God speaks from the cloud, "This is my beloved Son, with him I am well pleased":

> Now how could God pour out more of himself and offer himself in a more loving or sweeter way than by saying that his heart is pleased because his Son Christ speaks so friendly with me, is so heartily concerned about me, and suffers, dies, and does everything with such a great love for me. Do you not think that a human heart, should it actually feel that God is so well pleased with Christ when he serves us in this way, must shatter into a hundred thousand pieces because its joy is so great? For then it would peer into the depths of the fatherly heart, yes into the inexhaustible

---

84. Luther, WA X:I, 2, 297, cited by Althaus, *The Theology of Martin Luther*, 186.

85. Herein lies the importance of Lohse's statement (*Martin Luther's Theology*, 41) about the centrality of justification by faith in Luther's thinking: "the doctrine of justification is not something added to the question of the knowledge of God and the self. It must be taken up at the outset in thematizing such knowledge, if it is to be at all appropriately discussed."

goodness and eternal love of God which he feels and has felt toward us from eternity.[86]

In this way, in Luther's thought, the love of God in Christ, the Savior, is brought into closest proximity with the love of God in the Father, the Creator. As Luther observes in his characteristic, mystical language, about the Father:

> Where does a man who hopes in God end up except in his own nothingness? But when a man goes into nothingness, does he not merely return to that from which he came? Since he comes from God and his own non-being, it is to God that he returns when he returns to nothingness. For even though a man falls out of himself and out of all creation, it is impossible for him to fall out of God's hand, for all creation is surrounded by God's hand. So run through the world; but where are you running? Always into God's hand and lap.[87]

Luther envisions the "fatherly heart of God" with particular vividness in the activities of human work: "And so we find that all our labor is nothing more than the finding and collecting of God's gifts; it is quite unable to create or preserve anything."[88] Again: "God could easily give you grain and fruit without your plowing and planting. But he does not want to do so . . . But you are to plow and plant and then ask his blessing and pray; 'Now let God take over; now grant grain and fruit, dear Lord! Our plowing and planting will not do it. It is thy gift!'"[89]

In Christ, the believer thus discovers the "fatherly heart of God" everywhere, not just in salvation, but also in the whole visible creation. The Grace or the self-giving love of God, of the Son and of the Father, is thereby the theme that unites Luther's thought about salvation and creation in the grand narrative of God's works.[90]

---

86. Luther, WA XX, 228, cited in Althaus, *Theology of Martin Luther*, 187.
87. Luther, WA V, 168, cited in Althaus, *Theology of Martin Luther*, 111.
88. Luther, LW XLV, 327, cited in Althaus, *Theology of Martin Luther*, 109.
89. Luther, LW XIV, 114–15, cited in Althaus, *Theology of Martin Luther*, 109.
90. Cf. the summary statement by Gregersen, "Grace in Nature and History," 8: "Luther's theology of the Trinity is modeled after the pattern of the self-giving love of Jesus Christ. Not only the Son, but also 'the Father has given himself to us, with all his creatures' (Large Catechism, explanation of the First Article of the creed.) All this has been given me 'without any merit of worthiness on my part' (Small Catechism, explanation of the First Article). In a similar way, the Holy Spirit is conceived as the one who gives by letting all members of the church share in the community of God and saints . . .

That Luther understands the Spirit, the bond of love between the Son and the Father, also to be at work in salvation and creation is clear. That Luther also did not develop a theology of the Spirit that was commensurate in depth and vitality with his theologies of the Father and the Son also seems to be clear.[91] But that question cannot concern us here.

Seen from the perspective of Luther's Trinitarianism, then, we can identify how integral the whole creation was in Luther's theology, notwithstanding the fact that creation was at once the background of his thought. Driven by his existential situation to focus so much of his attention on salvation, and to see all things from that perspective, throughout his career as a reformer, and in that respect also driven to allow creation to drift to the background of his theological concerns, Luther nevertheless firmly believed that creation has its own integrity, and so argued, repeatedly and with real force.

This also helps to explain why Luther found it necessary, even against some of his own most heartfelt thoughts, to allow the sensibility of *seeing* to shape his thinking, sometimes in dramatic ways. To be *claimed by faith*, for Luther, one must shut one's eyes and listen only to the Word of God and, indeed, insofar as one's eyes are opened at that moment one can only see the ugly and contradictory specter of the Cross of Christ. But *to live in faith*, one can open one's eyes again, and see the glories and the beauties of the visible creation in a new way and encounter, in that seeing, the wondrous and overwhelming powers of the Creator, in, with, and under all things.

In this sense, for Luther, the believer can see not just the alienating mask of God in the whole visible creation, which hides God, but also signs of the overflowing goodness of the fatherly heart of God, working amazing blessings in all things. For Luther, then, the believer can contemplate the whole visible creation, in some measure, as Adam and Eve did before the fall and also see visions of a gloriously new creation of all things, a world that will be much more beautiful and transparent to the wonders of God's presence than the world of Adam and Eve was before the fall.

Alongside this outward-directed love, however, God is from eternity to eternity involved in the self-giving mutual love between Father, Son, and Spirit. The self-identity of God as love precedes and structures God's self-giving love toward creation."

91. See Gregersen, "Grace in Nature and History," 11.

For Further Discussion: The Motif of the Cosmic Christ

This investigation has concluded without addressing one of the most controversial elements in Luther's thought. Luther believed that, in some sense, the crucified, risen, and ascended Jesus Christ is *everywhere*, as God is everywhere—omnipresent or ubiquitous. I want to acknowledge the controversy about this theme in Luther's thought, here in conclusion, but I also want to underline what in my judgment is its promise. It has suggestive implications for further reflection about a topic that has in recent years emerged in theological discussions in the West with a new vitality, sparked by Joseph Sittler, whose thought I will consider in the next chapter—cosmic Christology.[92]

Some theologians have long ago concluded that Luther's ubiquity theology is a dramatic dead-end. Already Calvin, in the typical tonality of sixteenth century theological discourse, called Luther's thought about the ubiquitous Christ "monstrous."[93] Those who would *a priori* reject Luther's ubiquity theology—and anecdotally they are the vast majority of theologians who address such matters in our time, including numerous Lutherans among them—typically look at this theme in Luther's thought as an *aberration*, as an idea that he seized upon in the heat of the moment to pursue his polemics in sacramental theology against the likes of Zwingli.[94] Luther, they point out, never did much with the theme once the heated arguments of the sacramental controversies had cooled. But I believe that this theme is integral to his thinking.

What did Luther actually say about the ubiquity of the crucified, risen, and ascended Christ? This is one short answer.[95] The resurrected Christ ascends to the right hand of the Father, but not to sit there, as a stork sits on a nest, as Luther can colorfully observe, but in order to fill all things (Luther has in mind here one of his favorite texts, Ephesians 4:10: "He . . . ascended far above all the heavens, so that he might fill all things.") But how could this be?

---

92. See my review of these trends, Santmire, "So That He Might Fill All Things."
93. Calvin, *Institutes* 4.17.30.
94. This, essentially, is Lohse's reading of Luther in this regard.
95. An exhaustive and nuanced treatment of all these questions is available: Kleckley, *Omnes Creaturae*.

Yes, says Luther, Christ does sit at the right hand of God, but "the right hand of God is everywhere"—*dextera Dei ubique est*. Luther thus unites his ascension theology with his profound vision of the immanence of God in the whole visible creation, a theme that we have reviewed at length in this chapter. Since, then, the right hand of God is everywhere, the risen and ascended Christ, who is at the right hand of God, can really be present in, with, and under the bread and wine of the Eucharist, at any time and at any place, and indeed in many times and in many places, when he is so disclosed by the Word of God, in the power of the Spirit, to the faith of the people of God.[96]

The immediate concern of Luther's critics in this respect, and this is why Calvin called the idea "monstrous," is that Luther's doctrine seems to explode, as it were, the humanity of Christ. In terms from our own time: it would seem that the humanity of Christ, in Luther's view, is now, qua ubiquitous, akin to cosmic dust.[97] It is everywhere, it seems, but it is no longer recognizable as the Christ of the Gospels, the human Savior.

The issue here is what is called "the local presence of Christ" at the right hand of the Father. If there is no local presence, say the critics, there is no longer any human Jesus. Even the most astute and the most sympathetic of Luther's interpreters at this point, such as Kleckley, tend to agree. Kleckley grants that Luther, minimally, was not clear about this issue and that Luther said different things about it in different contexts.

Our best ecumenical theologians today, such as George Hunsinger, therefore want to downplay, if not totally eliminate, Luther's ubiquity theology.[98] Karl Barth, interestingly enough, notwithstanding his deep devotion to Calvin's theology, gave a qualified approval to Luther's ubiquity theology.[99] Obviously, we cannot resolve this issue here. But two points can be made, with a view to possibly getting this discussion underway again, with regard both to historical research and to theological reflection, precisely

---

96. So Luther says, LW LXIX, 221: "So, too, since Christ's humanity is at the right hand of God, and also is in all and above all things according to the nature of the divine right hand, you will not eat or drink him like the cabbage and soup on your table, unless he wills it . . . , unless he binds himself to you and summons you to a particular table by his Word, . . . bidding you to eat him. This he does in the Supper, saying, 'This is my body'"; quoted in Steinmetz, "Scripture and the Lord's Supper in Luther's Theology," 262.

97. Over against Luther's thinking, Calvin had already used the example of Christ's body being scattered in innumerable crumbs of bread. Calvin, *Institutes*, 4.17.

98. Hunsinger, "Aquinas, Luther, and Calvin," 181–93.

99. See Santmire, *Ritualizing Nature*, 283n17.

because this discussion has currently ground to something of an ecumenical halt.

First, a suggestion. It will be helpful for those who may wish to consider reopening the discussion about the ubiquitous Christ to revisit Luther's sophisticated paradoxical and mystical immanence theology. In a way, the whole ubiquity theme of Luther's thought stands or falls with his immanence theology. But sometimes the most learned of theologians shrug off Luther's immanence theology as if it had nothing to say, without considering its striking—even brilliant—ways of addressing both apophatic and kataphatic theological issues.

Is there a place for paradox in theology or not? And does the mystical tradition of the West, which fundamentally shaped Luther's views on these matters, still have anything to say to us? If, then, both paradox and mysticism *do* have a place at the theological table, shouldn't Luther's vision of the ubiquitous Christ be given a fair hearing in the conversations of contemporary theology?

Second, the issue of Christ's local presence at the right hand of the Father must be addressed and clarified. If the ascended Christ is now to be thought of as a kind of cosmic dust, then Calvin was right in his judgment, if not in his tonality. Any notion of the disintegration of the body of the ascended Christ is not acceptable. How can any Christian pray to a risen Christ whose body has, as it were, exploded?

Nor would it be acceptable, in a roughly parallel way, to allow discourse about the risen and ascended Christ to become, by default, if not by intention, discourse about some general cosmic force.[100] Pierre Teilhard de Chardin, who himself thought in terms of a cosmic Christic force, once asked himself, indeed, that very question: "Is this still the Christ of the Gospel?"[101] But consider two things, in this respect.

To begin with, we should acknowledge that in a sense all bets are off when it comes to talking about the risen and ascended Christ. No kind of literalism will work at this point (nor at any theological point, presumably). What indeed is a local presence in the context of modern cosmology? The thought of a local presence "above," as suggested in the Preface to the Eucharist, "Lift up your hearts," does not, by itself, make much sense. Where

---

100. This is the concern of Yeago, "Jesus of Nazareth and Cosmic Redemption," which leads him to stress the historical particularity of the incarnate Christ, in conversation with Maximus the Confessor.

101. Quoted in Yeago, "Jesus of Nazareth and Cosmic Redemption," 163.

is "up"? Further, local presence in common speech suggests something situated in a system of relationships between many other things, each of which is a local presence. What is a local presence—by itself, in eternity, as it were without any relationship to any other local presence, as in the case of the all-transcending Jesus Christ sitting at the right hand of God the Father?

Which is to suggest: the idea of a local presence is obviously a metaphor.[102] It may be a very useful and, in this case, a necessary metaphor, but it is a metaphor nevertheless. In this regard, indeed, it may be the case that Luther's use of multiple prepositions, as distinct from a single preposition, as the "up" of the Eucharistic Preface or the "in" of panentheism, is just more obviously metaphorical than alternatives that focus on a single preposition.

Still, with respect to the usage of the metaphor local presence itself, those who insist on the idea surely have a point. If believers are to give expression to their personal relationship with the historical, biblical Christ in the Eucharist and to pray to that Christ personally, such as in the Kyrie, "Lord have mercy," the humanity of the ascended Christ must be accorded its own, historically recognizable metaphorical integrity. There is no way for believers to think of themselves relating personally to something akin to clouds of cosmic dust or to some generalized cosmic force.

With these concerns in mind, then, what if we were to affirm unambiguously that the risen and ascended Christ does sit at the right hand of the Father *locally*, the same Lord who appeared to the disciples behind closed doors? What if we were to say, more particularly, that *the human nature* of the ascended Lord is *not ubiquitous*? Could we not then *also* simultaneously say—drawing on the rich and deep veins of Luther's paradoxical and mystical thinking—that "the right hand of God is everywhere"? This would allow us to affirm a doctrine of the local presence of Christ at the right hand of the Father, on the one hand, *and* the ubiquity of the right hand of God, on the other.[103]

---

102. One way to grasp the profound and pervasive functionality of metaphors in theological discourse is to revisit the pioneering study by McFague, *Metaphorical Theology*.

103 This approach stops short of affirming the "communication of attributes" (*communicatio idiomatum*) of the two natures of Christ, in a speculative, metaphysical sense. One reading of the latter idea is that the omnipresence of the infinite Divine nature is communicated to the finite human nature of Christ. That would mean that the human nature of Christ would then be infinite by that communication of attributes, hence,

Luther, after all, clearly did say that the right hand of God is everywhere, *dextera Dei ubique est*, while he was less clear about whether the human nature of Christ is ubiquitous. We should probably read him, then, in terms of his clearest utterances. In some statements, indeed, Luther seemed to advocate precisely the kind of understanding of this theme that I am suggesting here, that we think in terms of the local presence of the humanity of Christ and the ubiquity of the right hand of God: "Christ's body is everywhere because it is at the right hand of God which is everywhere, although we do not know how that occurs . . ."[104]

If this were to be considered a viable way to approach this matter, that, in turn, would have highly suggestive theological implications for our understanding of the relationship between the whole creation and human salvation, particularly as Luther envisioned that relationship. The Christ of salvation would then affirmatively and narratively be the Christ of creation, and vice versa. In the testimony of faith, more particularly, we could possibly tell of encountering the presence of Christ, in some manner, when we

---

according to this way of thinking, ubiquitous. Luther seems to have affirmed the doctrine of the *communicatio idiomatum*, however, not in a metaphysical sense (dominated by the conceptualities of finite and infinite), but in a more strictly theological sense. His main intention always appears to have been to accent the humanity of *the person of Christ* as the only way we know Christ as divine and human, and to avoid any suggestion that there is somehow something "more" to Christ than we know in Christ's humanity. Cf. Luther, LW XXII, 346: "The two natures are united in the single Person of Christ, and this Person is both God and man . . . [A]nd the attributes of each nature are imputed to the other." And: LW XXII, 352: "'For the communion of the natures also effects a communication of properties.' The ancient fathers diligently taught this and wrote about it."

For the purposes of this essay, the proposed construct is that the human nature of Christ is local and that the right hand of God is ubiquitous. Obviously, highly technical ideas about the incarnation of the Word are being touched on here; and, just as obviously, none of those questions can be resolved here. Cf. further, Althaus, *Theology of Martin Luther*, 193–98, and his comment about Luther's Christology: "His dogmatic theory which describes Christ as true God and true man is not unified within itself but displays contradictions." For an instructive, systematic review of issues embedded in the discussion of the *communicatio idiomatum*, see Ted Peters, *God—the World's Future*, 198–201, especially 199: "The driving direction of the Lutheran theology of the cross was toward the revelation of the sublime Godhead in the humble and humiliated humanity of Jesus."

104. Luther, LW XXXVII, 214, cited in Lohse, *Martin Luther's Theology*, 231. Cf. also, Luther, LW XXXVII, 69, cited in Steinmetz, "Luther and Loyola," 262 (italics added): "So, too, since *Christ's humanity is at the right hand of God, and also is in all and above all things according to the nature of the divine right hand,* you will not eat or drink him like the cabbage and soup on your table, unless he wills it. . . unless he binds himself to you and summons you to a particular table by his Word . . . bidding you to eat him. This he does in the Supper, saying, 'This is my body' . . ."

contemplate the lilies of the field or when we respond to the faces of the poor, as well as when we receive the bread and wine of the Eucharist.[105] Indeed, to adapt a phrase from Calvin, the whole visible creation might then be viewed, with the eyes of faith, as the theater of *Christ's* glory, even as an icon in the classical Orthodox sense.

This kind of thinking might also help to open the doors for us in the West to a new era of reflection about cosmic Christology, which, in turn, could have real promise for ecumenical conversations with those whose thought has been shaped by eastern Christian traditions. Eastern theology and spirituality, after all, is richly informed by the sensibility of seeing and, in particular, by grand icons of the cosmic Christ.[106] Such a dramatic turn in Western theological circles toward cosmic Christology might also have fruitful implications for the ongoing conversations between contemporary theologians and their dialogue partners in the natural sciences, especially those in the field of cosmological physics.[107]

As we consider such possibilities, finally, it will be helpful to remember that suggestive studies of cosmic Christology are already available in the West, beginning dramatically with the prophetic insights of Joseph Sittler, offered in 1961 to the World Council of Churches in New Delhi, and later developed by the imaginatively crafted systematic writings of Jurgen Moltmann.[108] Western theologians do not have to begin their discussions of the cosmic Christ *de novo*.[109]

As these discussions might unfold in the West, would it be possible then to imagine that *Luther* will one day be considered to be one of "the fathers" of these trends, of what would hopefully then be regarded as an era of richly flourishing discussions concerning the cosmic Christ? Be that as it may, I turn now to consider the thought of one of the pioneering modern champions of cosmic Christology, the aforementioned Joseph Sittler.

---

105. For a discussion of different ways of relating to the cosmic Christ, thus understood, see Santmire, *Ritualizing Nature*, 119–28.

106. For one Western appreciation of eastern traditions in this respect, see Yeago, "Jesus of Nazareth and Cosmic Redemption."

107. The discussions of the relations between theology and the sciences are enormous in quantity and varying in quality. For one good entry into this field, see George Murphy, *Cosmos in Light of the Cross*. For a comprehensive technical study, well-informed both theologically and scientifically, see Russell, *Cosmology from Alpha to Omega*.

108. For an introduction to these discussions, see Santmire, "So That He Might Fill All Things."

109. See especially the essays by a range of scholars in Gregersen, ed., *Incarnation: On the Scope and Depth of Christology*.

3

Joseph Sittler's Pioneering Vision of the Cosmic Christ

Nature Transfigured[1]

IF THE WHOLE VISIBLE creation—nature—was a vital and sometimes rich background in Luther's thought (Chapter 2), that background tended to fade from the picture as Luther's theological heirs claimed his theology for their own purposes in ensuing centuries. For those heirs, God and the human creature still occupied the foreground, sometimes dramatically, particularly regarding issues related to human salvation. The dominance of that foreground was virtually complete by the middle of the twentieth century, precisely at the time when Joseph Sittler (1904–1987) began his remarkable theological ministry.[2]

---

1. This chapter is a thoroughly revised version of my essay, "Reformation Theology of Nature Transfigured."

2. A number of ecotheologians have recognized Sittler as a pioneering theologian of nature. Jürgen Moltmann, for example, credits Sittler with initiating the whole modem discussion of theology and ecological concerns (Moltmann, *Way of Jesus Christ*, 276). Still, Sittler's contributions to this field unfortunately have generally been neglected until very recently.

This situation is now being rectified, since a selection from his works has been published, with a helpful editorial introduction and an analytical essay. See Sittler, *Evocations of Grace*. The secondary literature on Sittler is also growing. For a good short introduction, see Childs, "Nothing Less than Everything: Thoughts on a Sittler Legacy." For a general review of Sittler's theology, see Steven Bouma-Prediger, *Greening of Theology*; and Bakken, "Ecology of Grace." Sittler's historical setting and his historical significance has also been exhaustively and instructively studied in Pihkala, *Early Ecotheology and Joseph Sittler*. See finally the thoughtful short study by Saler: "Joseph Sittler and the Ecological

It would take us too far afield to detail that history from Luther to Sittler here.³ Suffice to say that, with the rise of modern industrial and technological society, particularly with the ascendancy of the mechanical view of nature and the emergence of the aggressive spirit of capitalism, the society at large and theologians in particular found themselves more and more distanced from nature by the middle years of the nineteenth century. That theological trend was further strengthened as nature came to be increasingly defined later in the same century as driven by a Darwinian struggle for existence. What did a loving personal God have to do with *that*?

Hence the task of theology, according to many of Luther's theological heirs, had to be radically theoanthropocentric. They understood the chief objects of theology to be God and humanity. As the Reformation tradition flowed through the nineteenth century, it was then clear to most, if not all theologians who claimed that tradition as their own, that theology would have very little to say about nature.

Luther's theological heirs, as a general rule, handed nature over to the scientists or to the entrepreneurs by default, sometimes to the poets or the painters. Meanwhile, in the theological schools, those who were writing in the Reformation tradition developed a new moral theology, influenced by Pietism's preoccupation with human subjectivity and shaped by the philosophy of Immanuel Kant, who himself had championed the subjective moral life—all set over against what was more and more perceived to be an alien world of nature.

The theology of God and humanity would become all the more rooted in the depths of the Reformation tradition in the first half of the twentieth century, as that tradition collided with—when it was not condoning or even celebrating—the rise of National Socialism. Adolf Hitler and his followers sought to establish a new human domination of the world by "the master race," on the basis of a "heroic" Nordic mythos of nature. And some German Lutheran theologians at the time seemed willing to lend credence to the whole Nazi mythos by espousing a new interest of their own in the theology of nature.

Against these trends, the Confessing Church rose up during the Nazi era opposing every "natural theology." The Confessing Church was driven by a rejuvenated theology of the Word of God. The confessors of

Role of Cultural Critique."

3. For a review of this theological period, see Santmire, *Travail of Nature*, ch. 7.

the Barmen Declaration shut their eyes on the claims of the sociospiritual reality of *their* time, as Luther had done in *his* time, and, again like Luther, listened to the Word of God afresh, and with radical devotion.

Call this the triumph of the sensibility of hearing in the modem Reformation tradition. Luther had championed hearing over seeing, but still had allowed the sensibility of seeing to shape a range of theological ideas, such as his rich understanding of the sanctified life with nature or his vivid vision of the New Heavens and the New Earth. But by the beginning of the Twentieth Century, Luther's limited but real dependence on the sensibility of seeing had virtually disappeared in the thought of Luther's theological heirs.

The existentialist theology of the Word of Rudolf Bultmann (1885–1976) and Karl Barth's (1886–1968) early theology of crisis and his later dogmatic exposition of the Word of God, as he began his monumental Church Dogmatics, are only two of the most prominent examples of the triumph of the sensibility of hearing in the Reformation tradition in the first half of the last century. This rigorously consistent theology of the Word, juxtaposed over against every theology of nature, was the dominant expression of Reformation thought up to the middle of the twentieth century—and remained so to a significant degree in ensuing decades.

In this theological milieu, anyone who would tamper with the theology of the Word, thus understood, particularly any Lutheran, risked calling forth a whirlwind of polemics, especially if there were any signs of a new kind of attention to nature. Any theological revisionist would therefore have to be prepared to face charges that he or she was facilitating the birth of a new theology of glory, of works-righteousness, even of subservience to demonic powers.

Sittler's Emergence as a Theologian of Nature

But Joseph Sittler saw things differently in the middle years of the twentieth century. Sittler felt comfortable, theologically, moving beyond—although not abandoning—the sensibility of hearing toward the sensibility of seeing, even though he had begun his public theological career by accenting the doctrine of the Word.[4] But this development in Sittler's thought would not become fully apparent until the later years of his theological trajectory, although it might be argued that he had presupposed it all along. Be that as

---

4. Sittler, *Doctrine of the Word*.

it may, Sittler's address to the World Council of Churches in 1961 presupposed a well-established engagement with the sensibility of seeing and the theology of nature.

This is how Sittler's theological story unfolded. Already in a 1954 (!) essay, "A Theology for Earth," Sittler announced what was to be for him a lifelong theological passion:

> The largest, most insistent, and most delicate task awaiting Christian theology is to articulate... a theology for nature as shall do justice to the vitalities of earth and hence correct a current theological naturalism which succeeds in speaking meaningfully of earth only at the cost of repudiating specifically Christian categories. Christian theology cannot advance this work along the line of an orthodoxy—neo or old—which celebrates the love of heaven in complete separation from man's loves in earth, which abstracts commitment to Christ from relevancy to those loyalties of earth that are elemental to being. (S 30)[5]

Right from the start, then, Sittler began to search for what he thought of as a third way, between and beyond theological naturalism and theological supernaturalism. Right from the start, in categories that I adopted later, Sittler had committed himself to be a "revisionist" in behalf of the theology of nature, to *critically reappropriate* the classical Christian tradition, as distinct from the other two main theological positions that were available to him at that time: to be an "apologist" for the received Christian tradition or to be a "reconstructionist," to seek to rebuild theology from the ground up.[6]

Pursuant to that revisionist end, in that same 1954 essay, Sittler identified two ways that humans had characteristically related to the world of nature in the past: (1) subsuming nature under human life and, (2) the exact opposite, subsuming human life under nature (S 27). In contrast, said Sittler (using the sexist language and categories that many of us, regrettably, employed in those days), "Christian theology, obedient to the biblical account of nature, has asserted a third possible relationship: that man ought

---

5. For the reader's convenience, I will cite my many references to the collection of Sittler's writings, *Evocations of Grace* (see note 2 above) in the text, e.g., (S 30).

6. Broadly speaking, a revisionist works critically with the received tradition; an apologist attempts to justify the received tradition; a reconstructionist commits himself or herself to building a new theology on new theological foundations. See my discussion in Santmire, *Nature Reborn*, 6–10.

to stand alongside nature as her cherishing brother, for she too is God's creation and bears God's image" (S 28).

Sittler then called attention to a text that we have already reviewed, Psalm 104, a doxolological parallel to the Genesis 1 creation account, as we have seen, and—as ecological theology began to emerge in the second half of the twentieth century as a field in its own right—a text that was to become a key rallying point for many:

> Here [in Psalm 104] is a holy naturalism, a matrix of Grace in which all things derive significance from their origin, and all things find fulfilment in praise. Man and nature live out their distinct but related lives in a complex that recalls the divine intention as that intention is symbolically related on the first page of the Bible. Man is placed, you will recall, in the garden of earth. This garden he is to tend as God's other creation—not to use as a godless warehouse or to rape as a tyrant. (S 28–29)

Sittler concluded that essay, suggestively, anticipating his 1961 New Delhi address, and even foreshadowing conclusions that have been drawn by others in twenty-first century theological discussions, concerning what is now sometimes called a "deep Incarnation":[7]

> The Incarnation has commonly received only that light which can be reflected backward upon it from Calvary. While, to be sure, these events cannot be separated without the impoverishment of the majesty of the history of redemption, it is nevertheless proper to suggest that our theological tendency to declare them only in their concerted meaning *at the point of fusion* tends to disqualify us to listen to the ontological-revelational overtones of the Incarnation. (S 31; italics original)

Already in 1954, then, Sittler had put in place the scope, if not all the content, of a theology that was predicated on what I am calling the theocosmocentric paradigm—to be contrasted with the theoanthropocentric paradigm taken for granted by Karl Barth and numerous other theologians of Barth's era. Sittler was already at that point announcing a theology of God and the whole creation and an ethos of human kinship with nature.

Notable also was Sittler's inclination to shape that theology, not solely or even primarily or in terms of a theology of creation and, with that, the frequently invoked stewardship ethic. Rather, Sittler thought primarily in

---

7. For an introduction to the deep incarnation discussion, see Peters, "Happy Danes and Deep Incarnation," 244–50; and Gregersen, "Deep Incarnation and Kenosis," 251–62.

terms of a *christological* vision. This would then allow him, even prompt him, with texts like Philippians 2:5ff. in mind,[8] to envision Christ as the Servant and, with that image, to propose an ethos of *service to nature*. This was a profoundly countercultural stance on Sittler's part. Few Christians in those days thought of the human relationship with nature in terms of service. Most, particularly those who took the responsible stewardship ethos for granted, preferred to think of the human relationship with nature in terms of management or mastery.

All this then came to a kind of consummate expression in Sittler's 1961 New Delhi address. Lifting up the claims of Colossians 1:15ff. in that presentation, Sittler forcefully set in place the theocosmocentric paradigm, with a vivid picture of what was to become his signature christological vision. Here is that text, in short compass, announcing the cosmic Christ (NRSV):

> He is the image of the invisible God, the firstborn of all creation; for in him all things in heaven and on earth were created, things visible and invisible... [A]ll things have been created through him and for him. He himself is before all things, and in him all things hold together... [A]nd through him God was pleased to reconcile to himself all things, whether on earth or in heaven, by making peace through the blood of his cross.

And this is Sittler's commentary, in part:

> These verses sing out their triumphant and alluring music between two huge and steady poles—"Christ" and "all things."... [A]ll things are permeable to his cosmic redemption because all things subsist in him. He comes to all things, not as a stranger, for he is the firstborn of all creation, and in him all things were created. He is not only the matrix and *prius* of all things; he is the intention, the fullness, and the integrity of all things: for all things were created through him and for him. (S 39)

That is Sittler's christological vision. This, then, for Sittler, is its fecund theological implication—with Sittler here anticipating Pope Francis' vision of nature as our common home:

---

8. In Philippians 2, Paul celebrated Christ Jesus, "who though he was in the form of God, did not regard equality with God as something to be exploited, but emptied himself taking the form of slave, . . . he humbled himself and became obedient to the point of death—even death on a cross" (vv. 5–8 NRSV).

> A doctrine of redemption is meaningful only when it swings within the larger orbit of a doctrine of creation. For God's creation can only be the cosmos which is man's home, his definitive place, the theater of his selfhood under God, in cooperation with his neighbor and in caring relationship with nature, his sister. (S 40)

The time is at hand, Sittler then announces, for a new "Christology of nature."

> We have had, and have, a Christology of the moral soul, a Christology of history, and, if not a Christology of the ontic, affirmations so huge as to fill the space marked out by ontological questions. But we do not [yet] have... a daring, penetrating, life-affirming Christology of nature. The theological magnificence of cosmic Christology lies, for the most part, still tightly folded in the Church's innermost heart and memory. (S 45–46)

In Sittler's view, Christians in the modern period lost the power of the robust faith in cosmic redemption articulated in the Letter to the Colossians, because they allowed the Enlightenment worldview to reign unchallenged. "A bit of God died," he says, "with each new natural conquest" (S 43). The claims of human autonomy ruled the day. The realm of Grace retreated.

This reign of human autonomy, Sittler continues, has left us and our world ecologically bankrupt. The relationship of humanity with its God-given home, nature, accordingly, has deteriorated profoundly in virtually every culture around the globe. Hence, Sittler concludes presciently, drawing on imagery from Colossians, "the root-pathos of our time is the struggle by the peoples of the world in many and various ways to find some principle, order, or power which shall be strong enough to contain the raging '... thrones, dominions, principalities' which restrict and ravage human life" (S 45).

So for Christians in our era, according to Sittler, the way forward must be through "a Christology expanded to its cosmic dimensions, made passionate by the pathos of this threatened Earth, and made ethical by the love and wrath of God" (S 48). Note the accent here on Christology rather than the global environmental crisis. Christians, in Sittler's view, should not be motivated to care for the earth, first and foremost, by the thought of any crisis, but rather by the reality of God's Grace, established cosmically in Jesus Christ, according to the witness of the Bible as a whole, of which the vision of Col 1:15ff. is but one prominent example:

> For it was said in the beginning that God beheld all things and declared them good, so it was uttered by an angel in the apocalypse of John, '. . . ascending from the east, having the seal of the living God: and he cried with a loud voice to the four angels, to whom it was given to hurt the earth and the sea, saying, Hurt not the earth neither the sea, nor the trees . . .' (Revelation 7:2–3 KJV) The care of the earth, the realm of nature as a theater of grace, the ordering of the thick material procedures that make available to or deprive men of bread and peace—these are Christological obediences before they are practical necessities. (S 48)

Nine years later, in 1970, in an essay entitled "Ecological Commitment as Theological Responsibility" (S 76–86), Sittler drew out implications of this vision for the Christian ethos. To this end, he forcefully announced the theme "the integrity of nature," one of the bedrock presuppositions of the theocosmocentric paradigm. Sittler did this in response to Jesus' saying, as Sittler believed it should have been translated, not "Consider," but "Behold the lilies of the field" (Matt 6:28):

> The word "behold" lies upon that which is beheld with a kind of tenderness which suggests that things in themselves have their own wondrous authenticity and integrity. I am called upon in such a saying not simply to 'look' at a nonself but to 'regard' things with a kind of spiritual honoring of the immaculate integrity of things which are not myself. (S 80)

Sittler argued that "this way of regarding things is an issue that the religious community must attend to before it gets to the more obvious moral, much less the procedural and pedagogical problems" (S 80).

This means, Sittler says, bringing into question the notion that humans in their historical experience and in their selfhood as individuals are so set apart from the rest of God's creation that they can deal with it in Olympian arrogance. We are in fact siblings of the whole creation, he concludes: we are called therefore to care for the creation. Which means, finally, in Sittler's view, "that ecology, that is, the actuality of the relational as constitutive of all our lives, is the only theater vast enough for a modern playing out of the doctrine of Grace" (S 85).

Sittler would then develop this core of theological reflections in a variety of directions in ensuing years. But we have seen enough thus far to have identified how creatively he claimed the theocosmocentric paradigm as his own, as early as 1954 and then with his grand public statement in 1961 and in the insightful argument of his 1970 article.

Sittler and the Sensibility of Seeing

Now I want to probe the dynamics of Sittler's thought more deeply. Somewhere beyond the midpoint of his little classic, *Essays on Nature and Grace*,[9] the careful reader can detect a subtle shift of sensibility. The accent of Sittler's discourse modulates from the act of *hearing* to the act of *seeing*. We have already encountered these theological dynamics in our review of Luthers theology of nature. The shift from accenting hearing was at least as important in Sittler's thinking. As this shift occurs in Sittler's case, we can witness the emergence of an imaginative theology of nature, freshly developed, while still affirming the tradition of Luther: a theology of nature transfigured.

Sittler begins his perambulatory reflections in that aforementioned, musefully argued volume with the announcement that the "rhetoric of grace" must be made congruent with the potentiality and the actuality of the "artifactual world of operations" characteristic of modern technological society.[10] The doctrine of Grace, he suggests, must be expanded to encompass the world of nature, now so sorely burdened and desecrated by an often mindless human exploitation.

"Might it be possible, along with such reinterpretation," Sittler asks, "to recover and release such an understanding and celebration of Grace as shall locate its presence and power and hope within man's life in the world-as-creation as well as within man's hearing and receiving of Grace through the Word and Sacrament of forgiveness?"[11] Then he proceeds—in traditional Reformation fashion—to listen anew to the witness of the Word of God, but in this case attending to how that witness is mediated by the letters to the Colossians and the Ephesians.

By the time the several streams of Sittler's meandering thoughts have come together fluently in his final chapter in *Essays on Nature and Grace*, the sensibility has noticeably changed. We encounter a shift in emphasis from hearing to seeing, from exploring and then expanding the biblical *rhetoric* of Grace to proclaiming and celebrating the biblical *vision* of Grace. "The

---

9. Sittler, *Essays on Nature and Grace*.
10. Sittler, *Essays on Nature and Grace*, 15.
11. Sittler, *Essays on Nature and Grace*, 17.

community of the people of God . . . ," he tells us, is "a people caught and held by a vision of a King, a kingdom, and a consummation and by the massive contexts of culture, history, nature, as fields of its holy disturbance."[12]

Sittler suggests that it is no accident that historic utopias have been formulated in the milieu of the Christian faith. Utopias have to do with a vision. Yet there is a decisive difference, he maintains: "Utopias owe their character and force to the vigor of the 'see what is possible!,'" while "the Christian vision" of what is possible "is engendered both by the realities of human existence and the promises of the God of its faith."[13]

This "vision of the New Creation is a product of [the] God who is affirmed in faith to be a Creator of the world, Redeemer of the world, and Sanctifier of the world."[14] Still, the critical point of interest here, as we reflect on Sittler's thought, is that, as he emphasizes, both utopian thinking and Christian theology project *visions* of reality.

A universal vision, indeed, is precisely what this world of "human crisis in ecology"[15] needs for the sake of its own survival and well-being, according to Sittler. As he explains:

> This effort to draw our regard for the world-as-nature into organic relation with the reality of divine grace is not a merely theoretical exercise; its intention is practical. Unless some huge, commanding, and primarily religious vision of the world's future can seize, release, and exalt our spirits so as to free us from our unregarding, arrogant, and ultimately suicidal operations with the creation, we shall continue to be bombarded by the awesome data of ecological disaster, while remaining bereft of a theological indicative adequate to the challenge, an ethic expansive enough to address the problem in its full dimensions.[16]

This shift from the *rhetoric* of Grace to the *vision* of Grace, from the sensibility of *hearing* to the sensibility of *seeing*, might not seem all that remarkable. For, one might ask, has not Sittler shifted the accent of his discussion precisely as the Scriptures do, when they direct our attention from

---

12. Sittler, *Essays on Nature and Grace*, 119.
13. Sittler, *Essays on Nature and Grace*, 5.
14. Sittler, *Essays on Nature and Grace*, 120.
15. Lutheran Church in America, *Human Crisis in Ecology* is a social statement and study guide by the Lutheran Church in America, edited by Franklin L. Jensen and Cedric W. Tilberg, to which Sittler contributed.
16. Sittler, *Essays on Nature and Grace*, 116.

this world to the next, from the valley of the shadows to the coming of the consummated Kingdom of God on the heights of Mount Zion? The Scriptures themselves evidently are shaped by this two-dimentsional sensibility of hearing and seeing.[17]

It begins with Genesis 1. The God depicted here inaugurates the whole economy of creation, redemption, and consummation by speaking the world into being, but not without also seeing, with every creative act, that the fruits of the Divine creativity are good and, indeed, taken as a whole, that they are "very good." Recall the almost liturgical repetition of the phrase "And God saw that it was good" in Genesis 1.

St. Paul also invokes the sensibilities of hearing and seeing in his own way. In this earthly pilgrimage we walk by faith, Paul says (2 Cor 5:7). And that faith, for him, depends totally on the preached word (Rom 10:14), so much so that Paul virtually identifies the two, as when he speaks of "the hearing of faith" (Gal 3:2). Still, there *is* seeing in this life, he seems to maintain: seeing with Christ as its spiritual object. Those who do not have blinded minds, Paul says, know the glory of God in the face of Christ (2 Cor 4:6).

But this seeing, as Paul understands it, has real limitations. In this life, we see through a glass darkly. Nevertheless, when the Last Day does arrive, according to Paul, then the people of faith will see perfectly, face to face (1 Cor 13:12). He urges his hearers, accordingly, not to look to "the things that are seen but to the things that are unseen" (2 Cor 4:18 NRSV), that is, to the things that are to come.

Perhaps the most powerful biblical expression of seeing faith is expressed christologically in the Gospel of John: "And the Word became flesh and dwelt among us, full of grace and truth; we have beheld his glory" (John 1:14)(RSV). We have beheld! Here seeing emerges at the center of the Christian faith, in the vision of the Incarnation. A more subtle, but no less compelling expression of the same motif is Jesus' testimony that God "clothes the grass of the field" (Sittler's translation): "*Behold* the lilies of the

---

17. Levenson, *Sinai and Zion*, has studied this biblical sensibility. He concludes (148–49) that "the dominance of ear over eye does seem to be characteristic of ancient Israelite sensibility. There is, however, one area in which the relationship is reversed, the Temple . . . [T]he Temple traditions, unlike the traditions of the history of redemption which are so much better known, are addressed to the eye of the viewer, not to the ear of the auditor."

field, how they grow; they neither toil nor spin; yet I tell you, even Solomon in all his glory was not clothed like one of these" (Matt 6:28-9).[18]

Such, in broad compass, is the biblical dialectic of hearing and seeing, as expressed from Genesis to the book of Revelation. Although seeing in its perfected form is reserved for the Eschaton—this theme is most apparent in the book of Revelation, where John the Seer envisions a new heaven and a new earth—and although the people of faith are first and foremost hearers of the Word in this life, they do, nevertheless, see the unseeable even now, in some measure, according to a range of biblical witnesses.

It seems clear, then, that Joseph Sittler's theology of nature was well-attuned in this respect to Scripture's attention to hearing and seeing. For biblical faith, as Sittler knew, the Grace of God has its own characteristic experiential dialectic: Grace elicits the *seeing* of faith, in some measure, as well as the *hearing* of faith.

To underline the significance of this observation, consider a commonplace phenomenological truth that Sittler instinctively took for granted: the *interconnectedness* of all things. This, of course, is the primary insight of all ecological thinking. Interconnectedness presents itself to us, however, in a variety of ways.

Hearing is the mode of interconnection, *par excellence*, between persons. People may touch or see one another—even smell one another. But, most characteristically perhaps, we *speak* with each other, even as we are interacting bodily. Thus dialogue is of the essence of authentic *human* interconnectedness, as Martin Buber showed classically in his prose poem, *I and Thou*.[19]

Seeing, on the other hand, is more characteristically the mode of existential interconnection—although not exclusively so—between human beings and all other earthly creatures. Even Buber, for all his deep pondering of his relationship with a tree, could not imagine the "I" speaking with the tree as a "Thou" *and* the tree responding in kind. The personal relationship with the tree, as Buber envisions it, remains poised at the threshold of speech. The same is true, Buber suggests, for our interconnectedness with some of the higher animals, even though relationships of this kind can be much more subtle. The personal relationship with the tree or the animals, in Buber's view, is sustained primarily by sight, a very special kind

---

18. Sittler, *Evocations of Grace*, 80.
19. Buber, *I and Thou*.

of sight, to be sure, a contemplative seeing charged with wonder, but sight nevertheless.

Now a biblically informed, existentially compelling theology of nature will, as a matter of course, identify and celebrate the interconnectedness of all things, of persons with persons, surely, but also of persons—both Divine and human—with the whole world of nature. Phenomenologically, seeing seems to bring with it an impetus toward the panoramic.[20] It would seem to follow, then, that a viable theology of nature will, as a matter of course, be predicated on the sensibility of seeing, as well as the sensibility of hearing, in order best to illuminate the interconnectedness of the Divine and human persons with nature, as well as human persons with each other, along with the interconnectedness between the so-called non-human creatures, who are sometimes referred to as otherkind.

A corollary of that observation is this: a theology predicated, exclusively or even chiefly, on the hearing of faith will, in all likelihood, be a theology that will *not* readily be able to offer a forceful account of the interconnectedness of *all* things and the biblical basis for that motif. If we are to articulate a persuasive ecological theology, in other words, the *rhetoric* of Grace will perforce have to find expression also in the *vision* of Grace. This, Joseph Sittler instinctively understood.

Theologians must *see*, as well as *hear*. Otherwise there will be no interconnectedness of thoroughgoing import, theologically speaking, between God or humans, on the one hand, and all the creatures of nature, on the other hand, as far as ordinary human engagement with them is concerned. Humans will know the natural world fundamentally not in its interconnectedness with God and humans, but in its disjunction, in its otherness, that is, its inarticulateness—which is to say, essentially as an alien world.

---

20. Cf. Jon Levenson, *Sinai and Zion*, 148, discussing Eric Auerbach's use of this kind of phenomenological analysis in reviewing the narrative of Abraham's near sacrifice of Isaac: "Auerbach's remark that in Genesis 22, 'the decisive points of the narrative alone are emphasized [in contrast to the Homeric epics that are filled with acute visual descriptions],' implies a worldview in which the ear dominated the eye, for it is characteristic of the ear to absorb only one message at a time . . . , to perceive sequentially, whereas the eye is capable of a *panorama* [italics original] in the etymological sense, 'the sight of everything.' The ear, which perceives meaning in the alteration between sound and silence, necessarily gives an account which is in Auerbach's term, 'fragmentary,' whereas the eye is able to sense an integral whole 'without lacunae.'"

The Sensibility of Hearing, the Sensibility of Seeing, and the Theology of Nature

That Joseph Sittler should call, variously, for a new theology of nature in the years after World War II, a theology that in some significant measure was rooted in a new, more comprehensive sensibility of seeing as well as hearing, and do so from within the tradition of Martin Luther—who at times could passionately disparage seeing, as we have seen—could have done nothing else than send shockwaves through the hearts of various Lutheran theologians and other students of the Reformation who first began to listen to what Sittler was saying. That, indeed, was the response of some leading theologians in Sittler's time, particularly those who were called, in those days, "neoorthodox," to Sittler's first publicly visible, and still his best known, statement of faith, his 1961 address to the World Council of Churches.

But what Sittler realized, with amazing foresight—and what many of his early critics apparently did not grasp—is that, with the emergence of the global environmental crisis that threatens the very survival of the human species, along with the lives of countless other species, the heroic epoch of a Reformation theology predicated chiefly on hearing faith had begun to lose its viability: not because of its theological substance, but because of its parochial scope.

Hearing faith certainly stood Luther in good stead when he confronted the papacy and those whom he called the enthusiasts. Hearing faith certainly stood neo-Lutheran thinkers, such as Rudolf Bultmann, in good stead, when they made incisive claims of faith to protect the integrity of personal existence in a world of sometimes rabid objectification. Hearing faith, likewise, certainly stood the confessors of the Barmen Declaration in good stead when they took their prophetic stand against Hitler.

But can hearing faith alone stand today's theologians of the Church in good stead as we face up to the enormities of the current environmental crisis and the ecological ravages of the creature who may best be described in this context, as Sittler often said, as *homo operator*? Is not something more required, some broader, more comprehensive theological horizon than that offered by the traditional theology of the Word?

It would be too much, surely, to claim that Joseph Sittler was another Martin Luther. But there is, at least, a formal parallel between the two thinkers that is worth pondering. In the midst of a profound societal upheaval, the global ecojustice crisis, which few theologians besides Sittler at the time of his first writings had the prescience to notice, this latter-day disciple of Luther did precisely what Luther himself had done in the Sixteenth Century and what the Barmen confessors had done in the first half of the Twentieth Century.

To invoke a figure I have already employed, Sittler shut his eyes. He shut out the social construction of reality current in his day, taken for granted by most theologians as well as by most citizens of the Western world at that time—and he listened to the Word of God anew, with radical devotion.

But the outcome of this listening differed from Luther's and from Barmen's. Tutored by texts from Romans and Galatians concerning the justification of the sinner, Luther tended to remain within the matrix of his *Urerlebnis* of hearing faith. For the most part, he never self-consciously sought to move beyond the premises that began with the hearing of faith and completed itself in his espousal of a theology of the Word of God—although he was enough of a Scripture scholar to have been tutored by biblical visions of creation's beauty, wonder, and eschatological promise, all of which are predicated on the sensibility of seeing.

The confessors of Barmen did much the same. One thinks, in particular, of Dietrich Bonhoeffer's devotion to Scripture-study at the Confessing Church seminary in Finkenwald, even as that community was living in a time of existential emergency. Adherence to the Word of God in that situation was a matter of life and death. Adherence to the Word of God for members of the Confessing Church was, for them, the cost of discipleship.

That way of thinking served Luther well in his time, surely, for a theology of the Word of God was itself capable of communicating the message of justification by faith alone. It was a theology, after all, for *persons,* especially for persons who were profoundly disturbed in their subjectivity. Likewise for Dietrich Bonhoeffer's leadership at Finkenwald. He lifted up the teachings of the Sermon on the Mount and the challenges of discipleship given with that segment of Jesus' teaching. This was what the Finkenwald community urgently needed, Bonhoeffer believed: to be obedient to those Words of God.

Sittler, in contrast, tutored by texts from Colossians and Ephesians concerning the unity of all things and the redemption of the whole universe through Jesus Christ, uncovered a new dimension, a more universal dimension, implicit within the matrix of his *Urerlebnis:* the dimension of seeing faith. In effect, although apparently not self-consciously, Sittler thus lifted up—and magnified—what was by his time had become a long-neglected theme of Luther's thought: a theme that might be called the sanctified seeing of faith. And in doing so, Sittler bequeathed to us a grand vision of nature transfigured.

Sittler and Nature Transfigured

In a variety of often poetic explorations, indeed, Sittler sought to envision a new world, as the circle of his hearing faith spiraled outward in every direction to become a universal *vision* projected by his seeing faith. Sittler sought to portray what he called "the sheer fecundity of the reality of Grace."[21]

Sittler heard in the Word about the "image of the invisible God, the first-born of all creation," how "in him . . . all things were created" (Col I: 15–16, NRSV). The Image! The hearing of faith thus elicited for Sittler a new, although surely not unprecedented, seeing of faith. The result was not a new theology of the Word of God alone, although Sittler always self-consciously sought to be a faithful student of the Word of God.

The result, rather, was more comprehensive. It was a theology of the Image of God, now understood not in terms of Genesis 1, but of Colossians 1, which is to say, christologically. As Sittler notes in *Essays on Nature and Grace,* he wanted *"to see the world* as the text speaks of it."[22] That he then proceeded to do, as he wrote and spoke creatively on many subjects, for most of his life. Colossians l became for him what Romans 1 had been for Luther and what the Sermon on the Mount had been for Bonhoeffer. And Sittler's theological interpretations brought upon him much polemical heat—as Luther's had occasioned in his era and likewise for Bonhoeffer, although, sadly, it was in Bonhoeffer's case polemics unto death.

And more. Given Sittler's accent on seeing faith, it was no accident that his insightful mind took him back, again and again, to the patristic teaching about Christ Pantocrator, that is, the figure of Christ as the ruler of all things.[23] This was the tradition, above all others, that envisioned Christ as the Image of God and, in turn, incorporated a reverence for visual images—icons—in its liturgy and spirituality, as well as its theology.

---

21. Sittler, *Essays on Nature and Grace,* 46.
22. Sittler, *Essays on Nature and Grace,* 42; italics added.
23. Sittler, *Essays on Nature and Grace,* 42.

Sittler recounts what we can think of as a kind of modest "tower experience" (*Turmerlebnis*) of his own, recalling Luther's, that is worth noting here. In the context of his discussion of Christ Pantocrator, he tells us: "In the Cathedral of the Holy Trinity at Zagorsk, Russia, during the Feast of the Dormition, standing for hours amidst the prayers of the faithful before the iconostasis with its Anton Rubleff icons—literal presences of the 'mighty cloud of witnesses'—I came to understand a mode of Christ's reality that shattered assumptions about Western christological comprehensiveness and beckoned toward partly forgotten dimensions of Catholic Christology."[24]

This was the way Sittler's theological program materialized, then. As Luther, having shut his eyes to listen to the Word in a time of crisis, sought to reconstruct his tormented inner world with his theology of the Word, and as the Barmen confessors did as Luther had done in their own situation of political crisis and subsequently sought to reconstruct the world in terms of discipleship, so Sittler, likewise shutting his eyes to listen to the Word in yet another time of crisis, sought to reconstruct what he perceived to be the threatened world of nature—with his theology of the Image

As he himself conceived the challenge before him, moreover, Sittler self-consciously sat at the feet of some modern seers of renown, whom he perceived to be responding to the same kind of challenge, among them Pablo Picasso. "Here is a man of fantastic endowment, of disciplined craftsmanship," Sittler tells us, "a man who wants to find a truth by probing the possibilities of things. But everything about our time has struck and penetrated and disorganized older ways of seeing and stating." Hence, Sittler says, Picasso is driven to dissolve the old order and reorder it: "He tears apart an accustomed appearance and represents a strange appearance. He disengages the structural components of the anatomy of figures and things and reorders them in fresh designs." Picasso desired, Sittler concludes, "to fashion a new order, to bring to being a new cosmology from a shattered one."[25]

Gerard Manly Hopkins was another such seer, who was, surely, Sittler's artistic mentor more than any other, because Hopkins's angle of vision, which likewise sought to fashion a new order and thereby to bring into being a new cosmology, was, unlike Picasso's, specifically Christian. Hopkins pursued his vision, indeed, with a self-conscious christological

---

24. Sittler, *Essays on Nature and Grace*, 55.
25. Sittler, *Anguish of Preaching*, 57.

intentionality of universal scope, which undoubtedly spoke to Sittler all the more.[26] With Hopkins's obedience to Grace, Sittler tells us, "all things became permeable to a new vision and demanded a holy evaluation."[27] Hopkins searched for a way "to set forth in outer reflection the inner nature of things," Sittler explains, "a quest for a language whereby a sense of the Grace of creation might be evoked for the reader."[28]

To articulate this quest, Sittler explains, Hopkins invented a new terminology: when the eye sinks into the surprise and the particularity and the interior gift-quality of things, we behold them as *inscape*. "Behold," Sittler observes—an affirmation that we have come upon already a number of times in these explorations—is precisely the proper word in this context, not "look" or "consider": "behold the lilies of the field" (Sittler's preferred translation, as we have seen).

In contemplating the inscape, Sittler continues, "one becomes aware of that energy of being by which all things are upheld, of that natural but ultimately supernatural stress which determines an inscape."[29] Steven Bouma-Prediger is undoubtedly right in observing that, for Sittler, the whole creation is a "sacramental mystery."[30] Every creature in this "theater of grace" (Sittler's own words[31]) is embraced by the energies of the cosmic Christ, in whom all things subsist with their own "wondrous authenticity" and "immaculate integrity."[32]

This is what the eyes of faith, focused by the witness of artists like Hopkins, allow us to see. So Sittler quotes Hopkins's own words from a letter: "I thought how, sadly, beauty of inscape was unknown and buried away from simple people and yet how near at hand it was if they had eyes to see it and it could be called out everywhere again."[33] This, it should be evident by now, was indeed Sittler's own richly nuanced, dramatically visual theological

---

26. Cf. the comment of Nathan A. Scott Jr., "Poetry and Theology of Earth," 112: "[for Hopkins,] the Incarnate Word or Christ is the real pattern on which all things are made . . . He is that first Inscape which is being adumbrated by all the inscapes of this world."

27. Sittler, *Anguish of Preaching*, 54.

28. *Anguish of Preaching*, 59.

29. *Anguish of Preaching*, 60.

30. Bouma-Prediger, "Conclusion," 229.

31. Sittler, *Essays on Nature and Grace*, 132.

32. Sittler, "Ecological Commitment as Theological Responsibility," 173 (cited in Bouma-Prediger, "Conclusion," 229).

33. Bouma-Prediger, "Conclusion," 229.

program: to show how near Grace is in all things, if only the faithful will have eyes to see, as well as ears to hear.

We should be very careful to identify the particularity of Sittler's theological program at this point. The hearing elicits the seeing. The seeing does not prepare the way for the hearing. Sittler's program is quite different from the agenda of traditional "natural theology." He eschews speculation. He commends the contemplation of faith. He eschews the logic of observation. He commends the logic of beholding.[34]

The difference is this, a point where Sittler self-consciously stands with Karl Barth, who also rejected the validity of any natural theology. For Sittler, there is a "single center" from which all the meanings, vitalities, and claims of theological reflection unfold. That single center is the historical particularity of a community "affirming itself to be a community of the Word of God—the Word as Creator, Redeemer, Sanctifier."[35]

This is his existential logic. Having shut our eyes in order to hear the Word in a time of crisis, we can then cautiously open them again to behold glimpses of a new cosmos, reconfigured in the Image of God, which is the universal inscape of the very Grace that first claimed us as we listened to and were claimed by the Word of God.

Sittler's theology of nature, this reconstruction of the world inspired by seers like Picasso and Hopkins, also calls to mind the spirituality of another seer in the Reformation tradition, one who looked to John Calvin as his "Church Father," Jonathan Edwards, who writes in his *Personal Narrative*:

> After this, my sense of divine things gradually increased, and became more and more lively, and had more of that inward sweetness. The appearance of everything was altered; there seemed to be, as it were, a calm, sweet cast, or appearance of divine glory, in almost everything. God's excellency, his wisdom, his purity and love, seemed to appear in everything; in the sun, moon, and stars; in the water, and all nature; which used greatly to fix my mind. I often used to sit and view the moon ... and in the day, spend much time in viewing the clouds and sky, to behold the sweet glory of God in these things; in the meantime singing forth with a low voice my contemplations of the Creator and Redeemer.[36]

---

34. "My theology is not one derived from nature, it is a theology of the incarnation applied to nature-which is quite different" (Sittler, *Gravity & Grace*, 67).

35. Sittler, *Gravity & Grace*, 63.

36. Quoted in Miles, *Image as Insight*, 2.

One could imagine coming upon such a passage in an as-yet-to-be unearthed journal of Joseph Sittler's adolescent or early seminary years. For, although Sittler's public utterances in his mature years adopted a much more restrained, urbane tone than Edwards's, Sittler's mature discourse surely leads us, by its many meandering courses, to the edge of the kind of vision of nature that so captivated the young Edwards.[37]

This is *a totally new theology of glory*, radically different from the one that Luther so vehemently rejected. It is not a theology of hubris, climbing up the ladder of human achievement or speculation toward a Divine throne, in order to establish, as the people of Babel once attempted to do, a tower of human power to be seen and applauded by all, and in order to be freed from the anguish and brokenness of this veil of tears.

This, rather, is a theology of humble faith, predicated on Jesus' Word to *behold*: a theology of kneeling down on the earth *(humus)* before the lilies of the field in gentle contemplation, a theology that has withdrawn any claims driven by will-to-power, a theology of waiting and watching and wondering in abject spiritual poverty in order to catch some sight of "the dearest freshness deep down things," in a world where all too often "things fall apart" and "the center cannot hold" and "the blood-dimmed tide is loosed."[38]

This is not a natural theology. This is a theology of nature. This is Joseph Sittler's legacy: the rhetoric of Grace *and* the vision of Grace, the invitation not only to hear, but, all the more so, to see. This is Sittler's Reformation theology of nature transfigured.[39]

Sittler's Invitation to See: Future Prospects

Can the ecumenical Church resolutely claim such a theology for itself today? Can the mind and heart of this Church itself likewise be transfigured?

---

37. Sittler's theology of nature might well have been enhanced had he found occasion to give Edwards's thought the same kind of attention he gave to the works of the Greek fathers and the Orthodox tradition more generally. Nature was one of Edwards's preoccupations, as was the construct of the image. On Edwards in this connection, see Jenson, *Americas Theologian*, 16.

38. Hopkins, "God's Grandeur"; Yeats, "Second Coming."

39. See, further, Santmire, "Toward a Christology of Nature." Later, I attempted to expand on Sittler's cosmic theology of grace, in particular, and cast it in more discursive terms, by building critically on the contributions of Jürgen Moltmann and Colin Gunton, in Santmire, "So That He Might Fill All Things."

Questions like these push us beyond the scope of this chapter, but they are exciting to contemplate briefly at the very end, as we ponder Sittler's invitation to see as well as to hear. Sittler himself drew some of these implications, at various points in his writings, as we have seen. Others remain to be articulated.

Consider the following sketches of such promising possibilities. I begin with some words of my own, concerning the Church's liturgy: "What one *sees* in the liturgy or in the Church building or while praying (must one always pray with eyes closed?) no longer can be regarded as a peripheral question or as 'adiaphora.' If the Church's liturgy is its mode of identity-formation, then the faithful must be encouraged to see glimpses of the glory of the Lord in each other's faces, in the sacramental elements, in the liturgical actions, and in the architectural shape and the imagery of the liturgical space."[40]

In this context, the visual arts will undoubtedly need to be much more consistently integrated into the life of the Churches of the Reformation today than they generally have been in the past. Can we imagine a day, for example, when works such as Van Gogh's "Potato Eaters" or "Starry Night" will be just as important in the life of American Protestantism as Bach's "B-Minor Mass" or Handel's "Messiah" are today? Is there a place, indeed, within Reformation Churches, *mirabile dictu*, for a new kind of biblically inspired iconography? Can the thought of "Reformation icons" be regarded any longer as a contradiction in terms?[41]

---

40. Santmire, *Ritualizing Nature*, 97. For further discussion of the need for liturgical reform in this ecological era, see Santmire, *Nature Reborn*, ch. 6; and a reshaped presentation of the same argument in Santmire, "Critical Challenge for Ecological Theology," 423–46. For a more extensive study of these issues, see my aforementioned book, *Ritualizing Nature*. Further, a place to begin all this, in practice, could be by lifting up and exploring the meanings of water in liturgical practices, in the context of renewed baptismal practices. For the latter, see the imaginative exploratory essay by Benjamin A. Stewart, "Stream, the Flood, the Spring," 160–76.

41. This is not to suggest that there has been no Protestant iconography. On the contrary, the popular religious impulse to address this spiritual need has taken over at the edges of Protestant life in the US and in the hearts of many American believers, too, as shown in the stunning study by David Morgan, *Visual Piety: A History and Theory of Popular Religious Images*. While Luther and Calvin urged allegiance to the Word and voiced suspicion (Luther) or rejection (Calvin) of images, generations of Lutheran and Reformed Christians in North America have taken images such as Warner Sallman's *Head of Christ*, *Christ at the Heart's Door*, and *The Lord Is My Shepherd* into their hearts and homes, contemplated them as illustrations in their Bibles, and sometimes even viewed them in church school classes and pastors' offices. The new Protestant iconography that Sittler's

This kind of concern for seeing would then have a place in the Church's moral praxis and public witness, too. For it is not enough to *tell* the faithful what to do or how to witness, and why. They must be *shown* by their lay and ordained leaders visibly, by the example of a tangible, enacted moral and spiritual vision, what God is calling them to do.

A case in point is the witness of the current Pope's most cherished saint, Francis of Assisi (1181/2–1226), and particularly St. Francis' creation of the living parable of the Christmas Creche, still so integral a part of popular Christmas piety in the West many centuries later. Can the faithful be taught to *see* there, in St. Francis' ministry in that forest setting, what St. Francis himself saw, when he gathered people from all classes and animals, too, around a rustic table for that Christmas Eucharist: the enactment of the eschatological Peaceable Kingdom that, in turn, calls into question the so-called wealth and so-called power of this world and that announces what Sittler called a new, holy obedience shaped by the poverty of that Child and by St. Francis' astoundingly humble love that flowed out to *all* creatures—not only to humans—as sisters and brothers?[42]

Then there is the believer's encounter with the larger world of nature itself, beyond the liturgical matrix, whether that is enacted in a cathedral or a forest. When with the eyes of faith I contemplate the fall foliage illuminated by the setting sun in the White Mountains of New Hampshire or when I walk around one of the great trees of Muir Woods in California in awestruck wonder or when I stop in my tracks to behold the lilies of some field somewhere, will I not then be in my right mind, and indeed be seeing things which my eyes cannot see by themselves, when I say under my breath—"Thank you, Jesus"?

Still standing before the larger world of nature itself now, beyond the liturgical matrix, I want introduce a theme that is new to this book at this point, and which I can only lightly sketch here. Thus far I have accented how the sensibility of hearing the Word can be richly enhanced by the sensibility of seeing the world, in particular nature, with the eyes of faith. Now

---

theology appears to me to mandate will have to be images informed by the heights and depths of the biblical vision and inspired by visionaries like Picasso and Hopkins. These images must profoundly transcend biblical and sentimental literalism yet also be self-authenticatingly powerful enough to connect with the needs of the popular soul, which all too often has had to rest content with creations by the likes of Sallman.

42. I have attempted to draw out some of the implications of Francis's life story for spiritual direction, in this respect, in Santmire, "Spirituality of Nature and the Poor."

I want to allude, further, to the promise of hearing nature with new ears, the ears of faith.

As we have witnessed thus far how the hearing of faith can, and hopefully will, whenever possible, give way to the seeing of faith, so consider this: that, inspired by the Spirit, we will on occasion be blessed by a *new kind of hearing* all over again—which can be called, according to Luther's way of thinking, a sanctified hearing. We will hear the songs and the groanings of the whole creation! This is that story, in very brief compass.[43]

Thanks to the ears of faith, I can stand silently to hear the forest in rural southwestern Maine singing and crying out in an early spring evening. Likewise, with the ears of faith, I can lift up my head, enchanted, to hear the birds beginning their songs at the break of a summer dawn at Mt. Auburn Cemetery near my apartment building in Massachusetts or, astounded, I can listen to the intense drumbeat of the thunder in any season, anywhere, hearing things which my ears cannot by themselves hear, and then say under my breath—"Thank you, Jesus?"

Likewise why wouldn't we choose to envision the same cosmic Christ working here on planet Earth in its anguish, struggling with the sometimes horrific facticity of evolutionary violence on this planet as it groans in travail? Likewise, more particularly, for the horrific facticity of historical human violence and the enormous pain that that violence typically causes as we humans join the whole creation groaning in travail? Can we not venture to envision the cosmic Christ here, too, in the midst of all this human groaning, all the more so perhaps, as the universal Suffering Servant?

At the same time, perhaps the more urgently because it is so difficult: do we not also need to try to envision with fresh theological imagination those aspects of the cosmos that we do *not* regularly encounter or never could see or hear directly or maybe would never want to see or hear, even if we could? Do we not need to envision the incarnate, crucified, risen, ascended, and omnipresent Lord, in, with, and under *all* things: as we are drawn into the—to us—alien world of wildness that Job taught us to encounter, the deadly darkness of cosmic black holes, for example, or, universally elsewhere, undetected by any immediate human knowing, the cosmic Christ drawing the whole cosmos toward its final end—toward both its catastrophic *finis* and its glorious *telos*, its ultimate termination and its ultimate fulfilment?[44]

---

43. See Santmire, "Two Voices of Nature."

44. To this end, we will be well-advised to begin to learn, perhaps for the first time,

Indeed, to adapt a phrase from Calvin: when the context is right, could not *the whole creation* be encountered, with the eyes and ears of faith as the theater of *Christ's* glory, even akin to an icon in the classical Eastern sense?[45] Recall Luther's words that all creatures are sacraments.

We can see and hear all this with good biblical warrant, too, using what Sittler once described as "the rhetoric of cosmic extension."[46] To this end, I have elsewhere explored how the Johannine and classical theological image of Christ as the Good Shepherd might also be interpreted *cosmically*.[47] If the "I am" statements of Jesus in the Gospel of John can be read as pronouncements of the creative Word of John 1:1, with Jesus here identifying himself with the God "I Am Who I Am" of the Old Testament, a not implausible thought, then the pronouncement "*I am* the good shepherd" in John can perhaps legitimately be read as having underlying cosmic meanings.

Imagine, then, this cosmic Shepherd bearing with all creatures, the ascended Lord, in, with, and under all things, suffering intimately especially with those creatures that experience pain, at times singling out some creaturely domains or even individuals for special care, as a good shepherd does, when leaving the flock behind and seeking out the sheep that is lost. Luther once suggested this kind of vocation for the cosmic Word, strikingly, when he commented on Hebrews 1:3, concerning the Word of God *upholding* all things. This is a Hebraism, Luther observed; it expresses "a certain tender and, so to speak, motherly care for the things which he created and which should be cherished . . ."[48]

The allusion to Luther here suggests one final thought, and this is a question for future research and reflection. As best as I can determine, Sittler never publicly wrestled with Luther's christological ubiquity theology. The latter, properly understood, as I suggested at the end of the previous chapter, might make it possible for Luther to be regarded as a "father" of cosmic Christology in the Reformation tradition. For Luther, as we saw, the

---

spiritual practices with the wilderness, in a Jobean mode (for Job, see my discussion above, Chapter 1, section vi) and perhaps to revisit the testimonies of wilderness writers like Thoreau and Muir. See, in this connection, the imaginative explorations in Dahill, "Rewilding Christian Spirituality," 177–96.

45. Calvin's theology of nature has been thoroughly researched and clearly described in Schreiner, *Theater of His Glory*.

46. In a personal conversation.

47. See my essays, referred to in n. 39.

48. Luther, LW XXIX, 112.

crucified, risen, and ascended Lord is everywhere, in, with, and under all things. But Sittler, a self-conscious Lutheran, appears to have by-passed this teaching of Luther altogether.

Absent further study, I can only speculate here about why this was the case. It appears to me that there may have been existential reasons why Sittler did this, perhaps reasons of which he was not even aware. From the beginning of his public theological career, Sittler's work was often attacked by a cadre of established Lutheran theologians, for a variety of reasons. Further, as we have seen, Luther's christological ubiquity theology itself has been—and still is—highly controverted, even in some Lutheran circles.

Could it be that Sittler, throughout his theological career, had too many theological irons in the fire to make it easy for him to add this one, too? I am not sure. But if Luther's christological ubiquity theology can be reclaimed as a significant historical contribution in the field of cosmic Christology, as I believe it can, perhaps those who treasure Sittler's theology will be able to find a way to draw on Luther's thought in this regard in order to enrich the kind of cosmic Christology that was one of Sittler's great gifts to the Church Catholic in our time.

4

The Theology of Nature as an Emergent Field of Promise

Multidisciplinary Reformation Explorations[1]

THE THEOLOGY OF NATURE is a relatively new movement in the world of Reformation thought and practice and therefore is neither widely understood nor easily defined, even by some who are variously involved in the movement.[2] But however one might understand this theological trend, this much we know. From the outset, particularly in the United States, Lutherans have been deeply involved, among them, as we have just seen, the theologian whom I—and numerous others—have ranked as a pioneer, Joseph Sittler. But he was by no means a solitary figure.

One might even argue that, as a group, American Lutherans have played a central role in the cultivation of this new field, both at the reflective, theological level and in the wider dimensions of Church life, especially by the production of two theologically substantive social teaching statements (1972, 1993) and by the emergence of a host of practical ministries

---

1. This chapter is a thoroughly revised paper prepared for the Convocation of the Association of Teaching Theologians, Evangelical Lutheran Church in America, at Columbus, Ohio, August 13–15, 2012. That paper, "American Lutherans Engage Ecological Theology," was published in the proceedings of the Convocation: Bohmbach and Hannan, eds., *Eco-Lutheranism*, 17–54.

2. One of the best recent attempts to do this is Jenkins, *Ecologies of Grace*.

in Lutheran circles that have embodied and, in some sense, tested the viability of the theological reflection and the social teaching statements.[3]

To be sure, as Hegel famously observed, the owl of Minerva does not take to flight until the dusk has come. Which is to suggest that it is difficult, if not impossible, to identify the meaning of any historical trend until it has run its course. It is much too early, I am suggesting, to draw any kind of satisfactory conclusions about where the theology of nature or ecological theology as a whole—now a global, ecumenical phenomenon—is going and what its influence might be, likewise for the particular significance of American Lutheran contributions to this field.

I will therefore restrict myself to historical impressions in this chapter, rather than trying to develop any kind of a comprehensive study. More

---

3. The focus on Lutheran engagement with ecological theology in this chapter is by no means intended to suggest, even implicitly, that Lutherans were the only ones during this period who were so engaged in the US. The development of ecological theology as a whole in the US, from its very beginnings in the middle of the twentieth century, was ecumenical in character.

Lutherans and Presbyterians, for example, worked closely together, from time to time, to identify an approach to this challenge. The Presbyterians also produced a substantive social teaching statement and a valuable theological guide of their own. The American Baptists, too, issued a statement on the environmental crisis in those early years. The Methodists, in turn, pioneered research and reflection about what they called (among the first, if not the first, groups anywhere to publicly identify this phenomenon), "environmental racism." The Methodist John Cobb was an early and forceful voice in ecological theology, likewise, and has continued to be so. See Cobb, *Is It Too Late?* Likewise, one of the classics of twentieth-century ecotheology was written by a Methodist: James Nash. See Nash, *Loving Nature*. A number of Catholics also participated in ecotheological conversations in the US in the last century, among them Denis Edwards, an Australian who published in America, works such as Edwards, *Jesus and the Cosmos* and Edwards, *God of Evolution*. Above all, perhaps, the theological influence of those who might cautiously be called the Catholic ecofeminists, figures such as Mary Daly (later ex-Catholic, indeed ex-Christian), Rosemary Radford Ruether, Sallie McFague, and Elizabeth Johnson—was and continues to be enormous. Any complete story of ecological theology in that era would have to tell their stories and identify their impact. It would also have to identify the kind of influence that global religious studies have had on Christian theology, particularly after the turn of this century. For a short account of the contributions of global religions in this respect, see Tucker and Grim, "Introduction: The Emerging Alliance of World Religions and Ecology," and the other essays in that *Daedalus* volume.

But sometimes it can be instructive to undertake the kind of vertical historical study of a single communion that I am pursuing here. I think of this, as I indicated at the outset, as studying a single tree in order better to understand the whole forest. When parallel vertical studies of the engagement of other Christian communions with ecological theology then become available, our grasp of the field as a whole from a horizontal, ecumenical perspective will hopefully be strengthened.

particularly, as I already noted in the Preface, I will explore the story of American Lutheran engagement with ecological theology as one who has had a hand, here and there, in shaping the first chapter of that engagement, for better or for worse.[4] This, of course, makes it all the more difficult for me to see the forest for the trees.

But this is what I think I know. The first chapter of American Lutheran engagement with ecological theology was not written by a committee, nor by any kind of "theological school" comprised of teacher or teachers and disciples. This chapter was written by a number of often isolated individuals who happened to have shared some theological and contextual assumptions and who were variously moved, some more self-consciously than others, by the challenge of fostering what a number of us thought of from the start as *an ecological reformation of Christianity*.[5]

I will endeavor, then, to identify some trends in this first chapter in the story of American Lutheran engagement with ecological theology, assess the significance of those trends as best I can, and then call attention to some areas that commend themselves for further discussion and field-testing, especially in the ranks of those Lutherans, but also others in the wider ecumenical world, who care about ecological theology and related ethical issues.

The beginning of the first chapter of the story I have in mind, in terms of historical significance, can be precisely dated.[6] In 1961, as we have already seen, a then little known American Lutheran theologian, Joseph Sittler, stepped to the podium of the World Council of Churches Assembly in New Delhi and delivered an address calling for a cosmic Christology. In

---

4. See my theological autobiography below, Chapter 5 (pp. 131-57).

5. In 1974, the staff of the Boston Industrial Mission and I organized an ecumenical conference at Wellesley College with the topic "An Ecological Reformation of Christianity?" This was a theme that suggested itself to numerous Protestants in those decades. See especially Nash, "Toward an Ecological Reformation of Christianity." That it is still a viable and, indeed, urgent theme has been more recently demonstrated by the authors of the essays in Dahill and Martin-Schramm, eds., *Eco-Reformation*; see especially the lead article by Rhoads, "Theology of Creation."

6. I am bypassing Paul Tillich. While I and others regard Tillich as a bona fide heir of the Reformation, Tillich himself, even as he recognized his indebtedness to Luther in particular, and to Protestantism more generally, did not regard himself as a Lutheran theologian, nor was he widely regarded as such by many theologians and practitioners in his own time. Nevertheless, in my view, Tillich's essay "Nature and Sacrament," in *The Protestant Era*, can be viewed as an important first step toward an authentically Lutheran ecological theology in the U.S., historically speaking.

retrospect that address can only be considered to have been a theological tour de force (so recognized, as I have already observed, by Jürgen Moltmann), although at the time many members of the then reigning theological guilds appeared to have had little or no awareness of what the import of Sittler's prophetic presentation actually was and therefore tended to downplay it to the point of insignificance or even derision. Sittler's address was published a year later. I will take the New Delhi address as the point of departure for this study.[7]

If 1961 is a clearly fixed point at which to begin these explorations, the end point of the first chapter of American Lutheran engagement with ecological theology is much more difficult to identify. We meet Minerva's owl once again. It is not easy to make judgments about the trends in which one is immersed and which have not run their courses. With total and perhaps entertaining arbitrariness, therefore, I will simply say this. The end of the first chapter of American Lutheran engagement with ecological theology is provided by the body of writings and practical ecclesial initiatives produced by a cadre of American Lutheran theologians and practitioners who, as of this writing, either have died or who are in or very near retirement. Which gives us a bit more than fifty years of theological engagement which to survey in the course of this investigation, a daunting task in itself.

Autobiographical Reflections: The Challenge of Cultivating a New Field

I begin with some autobiographical reflections (a matrix to which I will return, much more fully, in the next and last chapter of this book), in order to highlight the milieu in which those of us who were interested in an ecological reformation of Christianity initially worked and continued to work for some time. Today there is widespread awareness of the extent, if not the depth, of our global ecojustice crisis. Today is a time, more particularly, when the Christian Churches and their leaders around the world—Catholic, Orthodox, mainline Protestant, and Evangelical—have become highly visible advocates of ecojustice and when written works in ecological theology have proliferated virtually to the point of infinity.[8] All this is signaled by the global reach of Pope Francis' *Laudato Si'*.

---

7. Sittler, "Called to Unity."

8. It made sense in the 1990s for Peter W. Bakken, Joan Gibb Engel, and J. Ronald Engel to produce what was more or less a complete bibliography of English works in

The Theology of Nature as an Emergent Field of Promise

Those who live and work in this vast world of discourse may find it difficult even to imagine the theological situation faced by some of us who first addressed the challenge of an ecological reformation of Christianity in the sixties and seventies of the last century in the U.S. We knew that something momentous was unfolding in the world around us and we felt called upon to address the then emerging crisis theologically, but most of us also felt very much alone, without a viable theological support system.

When, for example, in 1963, I first broached the possibility of doing a doctoral dissertation on the theology of nature with my then recently assigned advisor at Harvard Divinity School, Gordon Kaufman, he told me that "theologians are no longer interested in nature." I remember those words vividly. Some years later, to be sure, Kaufman would do an about-face on this issue, and would become a highly vocal champion of the theology of nature.[9] But his 1963 comment to me typified the theological assumptions prevalent in those years in seminaries and graduate programs in theology, as well as in the preaching and teaching of the Church at the grass roots, at least according to the anecdotal evidence that I was able to accrue.

The theology we had inherited circa 1963 was self-consciously anthropocentric or, in Karl Barth's memorable language—which I have already deployed in this book—theoanthropocentric. Its chief concern was *God and humanity*, often to the disinterest in or even to the total abandonment of thought about the wider world of nature.[10] With the exception of only a few theological projects, such as Paul Tillich's[11] or, as we have seen, Joseph Sittler's,[12] dogmatic or systematic theology at that time was thoroughly theoanthropocentric.[13] Biblical studies, dominated by the self-consciously

---

ecological theology, *Ecology, Justice, and Faith: A Critical Guide to the Literature*. It would not make sense to produce such a printed work today, since it would be out of date before it made its way into readers' hands. Even some online bibliographical record of such works might not be all that helpful, since it would be difficult to keep up with the global sweep of such publications, printed or online.

9. Gordon Kaufman announced this about-face in Kaufman, "Problem for Theology: The Concept of Nature."

10. On Barth, see Santmire, *Travail of Nature*, ch. 8. This material is a summary of the findings of my doctoral dissertation, "Creation and Nature."

11. For Tillich, see Drummy, *Being and Earth*.

12. For Sittler, see the previous chapter.

13. An important exception to this rule is the group of theologians whom I have elsewhere called "reconstructionists," thinkers who generally held that the classical Christian tradition is ecologically bankrupt and who therefore concluded that Christian theology must be reconstructed from the ground up. Representative of this trend were the process

existential New Testament interpretation of Rudolf Bultmann and his followers, and by the over-against-nature Old Testament hermeneutics of G. Ernest Wright and the Albright school, were also generally theoanthropocentric. Christian ethics, whether domesticated in the form of personal, contextual ethics or more publicly responsible in the form of the ethics of technology or politics, was likewise mainly theoanthropocentric.[14]

We should not forget that there were profound contextual reasons behind this trend. It was not simply a matter, as it is sometimes portrayed, of anthropocentric arrogance, predicated perhaps on the spirit of Western imperialism. It was that in significant ways, but it was also—as I already have had occasion to observe—an expression of soul-shaking revulsion against the National Socialist ideology in Germany, and the "German Christian" movement in particular. The Nazis and their theologizing fellow travelers were champions of the theology of nature. Their ideology of *Blut und Boden* presupposed a heroic, amoral fascination with nature red in tooth and claw and a social Darwinian ethos of survival of the fittest. Not for nothing, then, did Barth sound his resounding *Nein* to Emil Brunner's proposal for a very modest, reconfigured natural theology.

Multidisciplinary Reformation Explorations in an Emergent Field of Promise

But during the second half of the last century numbers of Reformation theologians and practitioners in the U.S. did work energetically and imaginatively, some more self-consciously than others, in what was then becoming a new field, ecological theology. As a matter of course, moreover, they presupposed what I have been calling the theocosmocentric paradigm, a theology of God and the world (cosmos), no longer the theoanthropocentric paradigm, a theology of God and humanity, with the natural world interpreted in terms of the first two.

I have already identified Joseph Sittler's theology as theocosmocentric. Other key Reformation theologians in this era, who worked in a variety of fields, also adopted that way of thinking, as did two major Lutheran

---

thinker John B. Cobb Jr. (see *Is It Too Late?*) and the ecofeminist Rosemary Radford Ruether (see *New Woman, New Earth*). For the categorization of ecological theologians as "apologists," "revisionists," and "reconstructionists," see Santmire, *Nature Reborn*, 6–10.

14. For the first, see Fletcher, *Situational Ethics*; for the second, see Albrecht, ed., *The World Council of Churches Conference on Faith, Science, and Technology*.

social teaching statements and other practical ecclesial initiatives from this period, as I will now try to show.

Philip Hefner (1932–)[15] has focused much energy on explorations in *theological anthropology*, particularly as that field intersects at many points with the findings of the natural sciences, above all in his major 1993 study, *The Human Factor: Evolution, Culture, Religion*.[16] Hefner understands human beings to be thoroughly immersed in nature, especially in its evolutionary history, although distinct from other creatures in important ways. For Hefner, even though his chief interest is in the human being as created, co-creator, the primary objects of theological reflection are still God and the cosmos, which he understands to be an intricately interconnected, ecological whole.[17]

Behind all this, for Hefner, is the biblical vision of God's history with nature, announced in Scripture by the story of God's covenant with Noah and by the proclamation of the resurrection of Jesus Christ.[18] Seen from

---

15. Hefner was one of the authors of the *Christian Dogmatics*, edited by Carl E. Braaten and Robert W. Jenson, a work that I believe can be associated with theoanthropocentrism (see Santmire, "American Lutherans Engage Ecological Theology," 21–22). If I am right about that judgment, Hefner's place among most of the authors of the *Christian Dogmatics* would have to be considered to be ambiguous, since, in my judgment, Hefner's own thought is theocosmocentric, as I will now attempt to show.

16. Hefner, *Human Factor*. More recently, Hefner turned to explorations of the particular meanings of technology: *Technology and Human Becoming*. His interest in responding to the findings of the natural sciences has been longstanding. See, for example, his essay, "Can a Theology of Nature Be Coherent with Scientific Cosmology?"

17. In the following sketch of the doctrine of creation, Hefner essentially gives us a sketch of the theocosmocentric paradigm, without using that language: "The doctrine of creation not only serves as an essential framework on which the soteriological statements of faith depend for their credibility and meaning. It is also one of the chief resources for overcoming what has come to be known, perhaps exaggeratedly, as the 'unitarianism of the second article.' The object of concern in this phrase is a reduction of Christian theology to soteriology, which falsifies the Christian faith because it cuts off the larger connectedness between redemption in Christ and the panorama of God's intentions and actions from creation to consummation. Such a reduction also thereby cuts the link between redemption and the physical world, society, and world history. If theology does not overcome this tendency, it finds it difficult to relate the faith to such issues as ecological concerns, our vocation in society, and the manifestation of God's Spirit in the world's history" (Hefner, "The Creation," 272).

18. Consider this elegant theological testimony by Hefner, "Nature's History as Our History: A Proposal for Spirituality," 182–83: "[The Noah story] occurs as the second reading within the ritual of the Easter Vigil. As such, the rainbow covenant with Noah is connected to the resurrection of Jesus Christ. The meaning is unavoidable: the resurrection of Jesus of Nazareth is an event within a continuum of events in which God has been

any angle, then, Hefner's thought is fundamentally shaped by the theocosmocentric paradigm.

Ted Peters (1941–) has likewise been extensively interested in the interface—for him, the "consonance"—between science and theology, from the perspective of *systematic theology*. In this context, he, like Hefner, as a matter of course developed a keen interest in ecological theology and environmental justice over the years. He publicly entered the discussion of the environmental crisis in 1980 with theological themes that he would later bring to completion in a number of major works, such as: his vision of the eschatological fulcrum of theology, his view of God's consummating Future as encompassing the whole of cosmic history, not just human history, his understanding of the Church's vocation as a proleptic community called to embody the cosmic promise of God's Future, here and now, insofar as that is possible in a broken world, and his eagerness to engage secular thought, both in its popular and its most sophisticated expressions.[19]

By 1992, Peters' theological eschatology had come into full view, as what is perhaps his most important work, *God—the World's Future: Systematic Theology for a Postmodern Era*, shows.[20] Peters' vision of God, from beginning to ending, comprehends the whole cosmos and, above all, the cosmos' eschatological Future, not just God's history with humankind. For Peters, God creates and brings all things to fulfillment from that Future. God also initiates the consummation of the whole creation by sending Jesus

---

active, and the continuum includes the history of nature. Here God affirms a covenant with every living thing and with the earth itself, in full recognition that in light of the evil that is in the human heart, this sets up a lovers' triangle. In that triangle, consisting of God, humans, and all of the earth's other biological and physical systems, humans could find themselves outside the chainlink of the covenant with nature. God will never again permit that covenant to be breached in favor of humans at the expense of the earth . . . The rainbow covenant predicates God as a higher advocate for nonhuman nature.

"The proposals for organizing our consciousness contained in these packets of poetic, mythic information articulate themselves with forcefulness. They point outward in projecting possibilities for human involvement in community with the rest of nature that can make both for the wholeness of the Creator's covenant shalom and also for the terror that accompanies the destruction of that wholeness. Shalom comes when we consider that our calling to be: sibling to the geese and the spider; eye, tongue, and heart to sweet earth; covenant partner with earth and its birds, cattle and every beast."

19. Peters, *Fear, Faith, and the Future*.

20. Peters, *God—The World's Future*. For a more succinct treatment of the underlying themes of his thought, cf. Peters, *Science, Theology, and Ethics*, ch. 4: "God as the Future of Cosmic Creativity." He gives a full and focused statement of his eschatology in *Anticipating Omega*.

Christ to the here and now. Jesus, in Peters view, is thus the first embodiment or the prolepsis of the eschatological Peaceable Kingdom of the entire cosmos. Through Jesus, in turn, in Peters' view, God then calls together a community of the end-times to love the lost and to care for nature.[21] At its most fundamental level, then, Peters's thought presupposes the theocosmocentric paradigm, eschatologically elucidated, to be sure.

Terence Freitheim (1936–) has been one of several biblical scholars in the ecumenical community who have fostered a figure-ground reversal in *Old Testament studies* in recent years.[22] For this new reading of the Old Testament, the theology of creation, in general, and the theology of nature more particularly, is now the primary framework for biblical interpretation, rather than the theology of human redemption, as it was for the preceding generation of biblical scholars, such as the aforementioned work of G. Ernest Wright.

Freitheim's magisterial 2005 study, *God and World in the Old Testament: A Relational Theology of Creation*,[23] indicates the fruitfulness of a biblical scholarship that presupposes, consciously or unconsciously, the paradigm of theocosmocentrism. Fretheim shows, again and again, how the theologies woven into the Old Testament understand creation as one world, with which God has a history, and not as some alleged stage for God's history with humankind. More particularly, Fretheim documents how Gen 1:26–28, the notorious dominion text, is not to be read as an excuse for domination, as that biblical passage has often been understood, and how Gen 2:15 is rightly to be read in terms of Adam's serving and protecting the earth.[24] Fretheim's exegetical investigations thus, in effect, read

---

21. Cf. Peters, *God—The World's Future*, xii (italics original): "The exhilarating impact of the gospel is that it evokes in us the life of beatitude. In the Sermon on the Mount Jesus describes the life of beatitude as living a blessed life today in light of the coming of God's kingdom *tomorrow* . . . In the life of beatitude the Holy Spirit collapses time, so to speak, so that believers can share ahead of time in the oneness of all things that is yet to come . . . Amid the viciousness of devouring competition, one can envision the lion lying down with the lamb. Amid the desert of portending mass destruction, one can glimpse the river of life flowing from the throne of God. Amid the wanton lack of care for the beings and things of this world, one can feel the heart beat with the rhythms of the divine love that pervades and promises wholeness throughout creation."

22. In addition to Fretheim's work, two of the most important studies in this area are Hiebert, *Yahwist's Landscape*; and Brown, *Ethos of the Cosmos*. I leaned heavily on the works of each of these scholars in Chapter 1.

23. Fretheim, *God and World in the Old Testament*.

24. Fretheim, *God and World in the Old Testament*, 48–56.

the theocosmocentric paradigm as a fundamental datum of Old Testament theology.

In *liturgical studies*, a field that has often been self-consciously theo-anthropocentric in character, Gordon Lathrop (1939-) has interpreted the classical Christian liturgy as deeply embedded in God's good earth and indeed in the whole cosmos of God, both in the liturgy's various current formations and as profoundly shaped by eschatological hope. Lathrop's insightful 2003 study, *Holy Ground: A Liturgical Cosmology*,[25] shows how the formative ritual of the Christian community is developed not only as embedded in the earth and in the greater cosmos, but as a praxis that shapes the Christian life to be a way of caring for the whole creation, not just for other humans. To this end, Lathrop envisions the liturgy as rooted in local, natural places and as maximizing gratitude for the material gifts of God, such as water and soil.

Lathrop's work is also noteworthy because it is shaped throughout by a theology of the Cross—a hallmark, of course, of the Lutheran tradition since its inception—interpreted suggestively in terms of the Earth and the cosmos as a whole. Lathrop's achievement is perhaps all the more remarkable, because he takes the theocosmocentric paradigm as a datum, which, for him, requires no defense, as the subtitle of his major work, "a Liturgical Cosmology," indicates.

I mention here, too, in the midst of these explorations, my own work (1935-) in *historical studies* in the theology of nature, particularly my 2000 outline of classical Christian attitudes toward nature, *The Travail of Nature: The Ambiguous Ecological Promise of Christian Theology*.[26] I followed a method of motif-research in that work. I identified two major streams in the history of Christian thought about nature, the spiritual and the ecological (or what I elsewhere called the theology of ascent and the theology of descent). Those motifs are more or less expressions of the two paradigms that I am highlighting in this book, the theoanthropocentric and the theocosmocentric.

This method allowed me to explore the thought of a number of representative classical theologians[27] whose works were shaped by the ecological

---

25. Lathrop, *Holy Ground*. My own work in liturgical ecology, Santmire, *Ritualizing Nature*, is dependent on Lathrop's argument in significant ways.

26. Santmire, *Travail of Nature*. For overviews of my theological work, see Stewart, *Nature in Grace*, 39–88; Fowler, *The Greening of Protestant Thought*, 92–100; and Jorgenson, *Ecology of Vocation*, 103–9.

27. One of the several limitations of this book is that I did not attempt to review the

The Theology of Nature as an Emergent Field of Promise

motif. Ireneaus, the mature St. Augustine, St. Francis, Luther and Calvin, and Teilhard de Chardin can helpfully serve as conversation partners with those working in the field of ecological theology today. From this perspective, ecological theology in our era is *not something totally* new under the sun in the history of Christian thought (*pace* Lynn White Jr. and his many disciples). Something like the theocosmocentric paradigm has been presupposed by a major Christian theological tradition since the second century, if not before.

Larry Rasmussen (1939–) is the last theologian I want to feature here in my discussion of the first chapter of American Lutheran engagement with ecological issues. I will devote more time to his thought than I invested in studying the contributions of other theologians whose works I have discussed here, since Rasmussen's contributions are so broadly construed and so lucidly illustrative of the kind of theocosmocentric thinking that I have been accenting in this chapter and elsewhere in this book.

I read Rasmussen as a Lutheran theologian-ethicist, notwithstanding the fact that he has written publicly as an "ethical monotheist."[28] In my view, all along he has presupposed the kind of rejuvenated Reformation faith and praxis proposed by Dietrich Bonhoeffer's "secret discipline of faith" and by Bonhoeffer's view of "a world come of age."[29] That Rasmussen's theology has deep Reformation roots will be apparent, once we attend to his theological argument.

Not for nothing does the theology of the Cross emerge organically in the flow of Rasmussen's exposition in his major, prize-winning 1997 study, *Earth Community, Earth Ethics*.[30] Not for nothing, as well, did he choose to write from the context of communities of creative ecojustice formation, especially those, such as a variety of Christian communities that have considered themselves to have been eschatologically shaped by the biblical vison of the coming Peaceable Kingdom, announced by Jesus. Not for nothing, likewise, has Rasmussen celebrated Luther's often misunderstood vision of the divine immanence, of God "in, with, and under" the whole cosmos. For

---

classical mystical traditions of Christian life and thought regarding nature, a context in which numerous women theologians flourished.

28. Rasmussen, *Earth Community, Earth Ethics*, xiii. For convenience's sake, I will hereafter cite this work in the text and the notes as "R" followed by a page number: e.g., (R 10.)

29. For Rasmussen's relationship to Bonhoeffer, cf. de Gruchy, "Concrete Ethic of the Cross"; in Spencer and Martin-Schramm, eds., "Fidelity to Earth."

30. See n. 28 above.

Rasmussen, above all, in historic Reformation fashion, everything depends on Grace, and then faith.[31] Everything, in Rasmussen's view, particularly the biophysical matrix in which we all live and which lives in us, is a Divine gift. Everything, in this sense, for Rasmussen is sacramental.

On the other hand, Rasmussen gives shape to that vision not in the evocative christological terms of a Joseph Sittler, but with his own compelling pneumatalogical and sacramental musings, terms that some Lutherans and other Protestant fellow-travelers might, perhaps ironically, find to be new and therefore not immediately accessible.[32] Still, in my view, Rasmussen moves freely within that Reformation theological movement that I have been highlighting, presupposing throughout the theocosmocentric paradigm.[33]

31. See (R 349–54).

32. And not only Protestants. Consider this response to Rasmussen's *opus magnus* by Charles R. Pinches ("Eco-minded"), chair of the theology department at the Jesuit-run University of Scranton, in the *Christian Century* (August 12–19, 1998) 756: "[O]ne . . . feels the nagging tension [in Rasmussen's book] between explicitly theological categories and those of the environmental crisis and deep ecology. In Rasmussen's case, however, this is not because theological concepts are radically revised [as is done by the advocates of deep ecology] but because theology is not the primary language of the book. While Rasmussen does a bit of theology here and there, the book lacks a theological structure. He never decides to consider systematically or historically what Christian theology has to say about ecology, the earth, or even creation. As a result, there is no theological context into which the reader can place the book's otherwise quite interesting reflections about our environmental troubles." Which is damning, I suppose, by faint praise.

It is of particular interest to me in this connection, that one of the frequent criticisms that was directed against the works of Joseph Sittler was that, in a sense, he quoted poetry too much, and that he did not identify the foundations of his theological argument with sufficient clarity. Could it be the case that there is something about thinking under the influence of the theocosmocentric paradigm that requires us to plumb meanings from the arts and the sciences, as well as from the normative theological tradition?

33. I do not want to push the question about Rasmussen's Reformation credentials too far. On the one hand, who cares? The theological/ethical challenge before us in today's world is too great for such parochial-sounding questions. On the other hand, *I* care, since I am instancing him and his work under the rubric of American Lutheran engagement with ecological theology. So it has been necessary for me to say something about this question, in order not to say nothing, especially given the character of what might be called Rasmussen's post-Lutheran Lutheranism. If anyone would like to pursue this question further, I refer him or her to one of Rasmussen's former students, Cynthia Moe-Lobeda, "Christian Ethics toward Earth-Honoring Faiths," 146: "Larry's deeply critical, appreciative, and reconstructive relationship with Lutheran traditions, in which he stands, demonstrates his approach to tradition. His Earth ethic and call for eco-Reformation are notably Lutheran . . . [He] finds riches in central Lutheran theological themes; and 'thinks creatively with' Luther, Bonhoeffer, and, to a lesser extent, Joseph Sittler and

Rasmussen's thought can be approached from many different angles, as a recent Festschrift in his honor revealed.[34] I will concentrate here on those elements of his theology that disclose his particular expression of a paradigmatic theocosmocentrism. And I will interpret Rasmussen's thought mainly as it came to expression in his volume *Earth Community*, already referred to, although a complete exposition of his thought would also have to deal at length with his more recent, prize-winning volume, *Earth-Honoring Faith: Religious Ethics in a New Key*,[35] as well as other works.

All that Rasmussen writes in *Earth Community* depends on his analysis of the current global ecojustice crisis, which he describes vividly in cultural, social, and scientific terms, yet with an underlying theological *cantus firmus*.[36] His search, he says, is "for an earth cosmology and an earth ethic, carried out in the recognition that nature and earth compose a single community. Whether we like it or not, it's life together now or not at all. Earth faith and earth community—this is humanity's next journey" (R 19). Rasmussen insists that we are concerned here with one community, indeed, "not of culture and nature, or history and nature, but of culture and history in and as nature" (R 32). This is the kind of vision of the one created world that the theocosmocentric paradigm fosters.[37] This is the vision of the one created world, which as a whole has its own integrity and each part of which also has its own integrity (R 98).

Rasmussen develops his argument in *Earth Community* with a kind of inductive sensibility, rather than beginning, say, as Sittler did, by exploring and explicating the meaning of key Christian symbols. Rasmussen's point of departure is the world as all can in principle know it, seen globally and through the eyes of numerous cultures. This is the global context in which the theological *cantus firmus* can be heard, he believes. People of all cultures cannot resist speaking of religious concerns, and dreaming dreams and seeing visions. "Whatever the wishes of the cultured despisers of religion," Rasmussen comments, "as a species we yearn to see things whole

---

other Lutheran theologians to retrieve and reconstruct less recognized resources proffered by their work."

34. See n. 29 above.

35. Rassmussen, *Earth-Honoring Faith*.

36. Since I will cite Rasmussen's *Earth Community* often, I will do so in the text, e.g., (R 26), for the reader's convenience.

37. Cf. (R xii): "The world around us is also within. We are an expression of it; it is an expression of us."

and sacred. We insist on telling a cosmic narrative and locating ourselves somewhere in it" (R 178).

In this global religious context, in Rasmussen's view, "the peoples of the Book," Islam, Judaism, and Christianity, hewed a bold line from the beginning that Rasmussen wants to commend, "a certain focus and concentration on community and social justice as a God-given vocation" (R183). Rasmussen thus wants to commend these particular religious traditions, which have kept faith and justice-ethics in the closest possible relationship. But not at the expense of a vibrant faith itself.

So he argues that "an evolutionary sacramentalist cosmology offers the richest conceptual resources for addressing earth's distress," on the one hand, but that that cosmology must be "infused with a profound earth asceticism and married to prophetic efforts aimed at 'the liberation of life: from the cell to the community' [Charles Birch, John Cobb]," on the other (R 247).

Seamlessly, then, Rasmussen moves from his general discussion of the human need for a cosmic narrative, which need, in his view, religions address, and from his commendation of religions of the book, in particular, for their accent on communities of social justice, to explore the ambiguities and the promise of the Christian tradition more particularly.

First, regarding the ambiguities of the Christian tradition: Rasmussen presents a nuanced critique of the classical Christian accent on contempt for the world (*contemptus mundi*). Yes, wealthy Christians, in particular, must be claimed by a new ascetic spirit today, he says. But all too often, in Rasmussen's view, the attitude of contempt has more generally paved the way for rapacious patterns of hostility against the poor, women, and the whole Earth, as well. Even in our own era, he says, "neither existentialism, neoorthodoxy, liberalism, common church practice, nor society at large in the North Atlantic world has a cosmology worthy of the name in many influential circles" (R191).

In this connection, Rasmussen mounts an extensive and insightful critique of the familiar American Christian fascination with stewardship (R 230ff.). He illustrates these dynamics with a revealing account of his own efforts, which ultimately failed, to influence the 1991 Canberra meeting of the World Council of Churches to move beyond stewardship theology (R 227–28).

Such a theology, in Rasmussen's view, will inevitably prove itself to have been counter-productive. It all too easily goes with the flow of our freewheeling industrial society. On the contrary, he believes, what is needed

The Theology of Nature as an Emergent Field of Promise

now is for Christians, and adherents of other religions and ways of life, too, to claim or reclaim "those symbols that effect a 'reenchantment of the world' [Max Weber] that edges out the deadly cosmology of mindless and valueless nature worked over by ghostly human freedom in all too much of modernity" (R 194).

What can the Christian faith—or the faith of many "Christianities," as Rasmussen prefers to say—do to respond to this situation which requires a new and enchanted global cosmology, which can heal the earth and transform the Christian life? Something radically new is required, Rasmussen announces, as he quotes the cry of the Korean theologian, Chung Hyun Kyung, on the floor at the Canberra meeting, with reference to the then emerging voices of the so-called Two-Thirds World: "We are new wine. You will not put us in old wineskins" (R 233). Rasmussen then tries to suggest some directions, perhaps not yet fully developed proposals, he allows, for further theological reflection.

To this end, he reaches deeply into the currents of the Lutheran tradition, as an exercise in critical, but creative theological ressourcement. He explores Luther's rich theological immanentalism, in particular. Luther presents us, in Rasmussen's view, with a cosmos—not just human history—charged with the presence of God (R 272–73). The whole cosmos is God's, intimately, powerfully, and pervasively, in Luther's view. Luther further accents the solidarity of humankind with otherkind, with the animals, in particular. Tutored by Luther, in a word, the universe can once again be enchanted for us. And we can care for nature in solidarity with all the creatures of nature. This is Rasmussen, summarizing that aspect of Luther's vision:

> [Luther's] *finitum capax infiniti*—the finite bears the infinite—is grassroots earth theology. It is earthbound and limited. That is God's way, among us. The body, nature, is the end of God's path. God is not a separate item, even a very large one, on an inventory of the universe, but the universe itself is God's "body" . . . God is not totally encompassed by the creaturely, but the creaturely is the one and only place we know the divine fullness in the manner appropriate to our own fullness. Experiencing the gracious God means, then, falling in love with earth and sticking around, staying home, imagining God in the way we can as the kind of creatures we are. The only viable earth faith is thus a biospiritual one. Earth ethics is a matter of turning and returning to our senses. The totality of nature is the theater of grace. The love of God, like any genuine love, is tactile. (R 280–81)[38]

38. I disagree with Rasmussen's interpretation of Luther at this point. Nowhere that

Celebrating Nature by Faith

The second major theme from Luther's theology that Rasmussen commends to us to consider as we seek to identify a new cosmology appropriate to our own times of global crisis is Luther's passionately affirmed theology of the Cross (R 282–83). This means for Rasmussen, to begin with, drawing on Luther's own images, that

> this Jesus is wholly of earth. He is not a fleeting docetic visitor, nor a ghostly bearer of gnostic truth, but really mortal flesh and blood from the countryside. Joseph tickles his bare belly button and covers his bare bottom; Mary puts his hungry mouth to her bare breast. (R 283)

It also means, for Rasmussen, a dual reference, to the universe as whole and to the meaning of this one particular person, Jesus:

> Yes, God is the ultimate life-source of the entire universe, its creator, sustainer, redeemer; and this God is disclosed in the cosmos as a whole. But, in the manner appropriate to human experience and knowing, this life-source is disclosed most compellingly in Jesus. This Jesus is the incandescence of God in human form. (R 282)

And this Jesus, born in and of the earth, is then made known by the formation of a people whose mission is to display redeemed creation as a just community. "Such is the pattern for both the formation of Israel and the 'People of the Way' of Jesus (Acts 4:32–35)," Rasmussen explains. "This is Luther's argument for Jesus as the masked clue to the revelation of the Ineffable One. A humanly experienced historical event opens onto an apprehension of all reality" (R 283).

But Rasmussen believes that perhaps Luther's most profound contribution to aid us in our quest for a new and vital cosmology is Luther's sometimes misunderstood view of the suffering of Christ. "What is discovered via Jesus," Rasmussen says of Luther's perspective, is this:

> only that which has undergone all can overcome all. In this sense, cross and resurrection ethics is an utterly practical necessity. Suffering, in its many expressions among its many creatures, will not be redemptively addressed apart from some manner and degree of angry, compassionate entry into its reality, some empowerment

---

I am aware of did Luther suggest that the world is God's "body." For a discussion of this, and related, issues in Luther-interpretation, see Chapter 2 above. On the other hand, in every other respect, I believe, Rasmussen presents a balanced and instructive interpretation of Luther's thought.

from the inside out, some experience of suffering as both a burden and a burden to be thrown off, some deep awareness of it as unhealed but not unhealable. (R 286)

Rasmussen observes that Lutheran Cross and Resurrection theology is thus curiously optimistic. It has seen the worst and discovered a mighty power for life. And this leads to a profound ethic of compassion and solidarity that "seeks out the places of oppressive suffering in order to overcome suffering's demonic, or disintegrative, manifestations . . . Its quest is not for victims but for the empowerment needed to negate the negations that generate victims . . . It insists that environmental justice is also social justice and that all efforts to save the planet begin with hearing the cry of the people and the cry of the earth together" (R 291).

Rasmussen has many other things to say in explicating his vision of an Earth ethics for an Earth community, both in *Earth Community* and in a number of his other writings. But we have seen enough at this point to permit this judgment, that this Reformation theologian and ethicist has dreamed dreams and seen visions of God and the whole creation, surely not just God and human history, in a way that makes it possible for us to see, if we are so inclined, the whole world as reenchanted with the presence of God and also God's incandescent and compassionate self-disclosure in that person of the earth, Jesus and his Cross, with Jesus thus pioneering the way of suffering love for the whole creation as it groans in travail.

In that vision, in Rasmussen's terms, we may also see an ethos of deep caring for every creature transfigured into the struggles of justice, again, for every creature, especially for those who suffer and are oppressed, a struggle to be claimed by that community of ecojustice that has received the name of that very Jesus for the sake of the whole world.

Practical Ministries in an Emergent Field of Hope

While such a list of theologians, in addition to Sittler, shows that the new field of ecological theology in American Lutheranism was a significant historical trend during the second half of the last century and beyond, there were also more practical expressions of ecological theology in Lutheran life and thought during the same period, that had their own import. The theology of the Church, it probably needs to be observed, is not exclusively the province of the Church's professional theologians. In this spirit, I want to call attention to two significant Lutheran social teaching statements and also highlight the emergence of some impressive on-the-ground ecotheological ministries.[39]

The 1972 statement by the Lutheran Church in America was called *The Human Crisis in Ecology*, as was the theological guidebook that was circulated publicly with the statement during its approval process.[40] Of special interest for my purposes here is the central theological chapter of the guidebook, "The World as Community." Not only did that chapter speak of God and the whole creation, nature included, as a community, it also highlighted what it called "the integrity of nature."[41]

Moreover, both the guidebook and the statement itself favored the language of caring, rather than stewardship language (that term was used only once in that 1972 statement) in their references to the human–nature relationship when it is as it should be. Both the statement and the guidebook also stressed the importance of social justice in response to all ecological concerns.[42]

Later, the 1993 social teaching statement by the Evangelical Lutheran Church in America, "Caring for Creation: Vision, Hope, and Justice" (which was not published with a guidebook), was, in my judgment, a much sharper and richer document.[43] Its biblical foundations were more clearly identified and its analysis of the then current crisis was much more extensive than the 1972 initiative, and, likewise, its ethical discourse was at once much more comprehensive and much more sophisticated.

---

39. A more complete treatment of Lutheran social statements in this era would also review a 1970 statement on the environment by the American Lutheran Church (ALC). I am not considering that statement in this chapter for two reasons: first, to keep my own discussion within reasonable limits; second, because the LCA statement was accompanied by a study guide, which set forth underlying theological understandings explicitly. The ALC statement is available in the archives of the Evangelical Lutheran Church in America: https://www.elca.org/WhoWeAre/History/ELCAArchives/ArchivalDocuments/PredecessorBodyStatements/AmericanLutheranChurch/TheEnvironmentCrisis.aspx/.

40. Jensen and Tilberg, eds., *Human Crisis in Ecology*.

41. Full disclosure: I wrote that chapter (and another) and also helped—as a member of an interdisciplinary team, which included Joseph Sittler—to draft the statement itself. Sittler wrote the concluding chapter.

42. A further popular publication came out of the working group that produced the 1972 statement, cowritten by the chair of the group, who was an academic biologist, and myself: Lutz and Santmire, *Ecological Renewal*.

43. Full disclosure: I also helped to draft this statement, as one member of an interdisciplinary team.

But like the 1972 statement and guidebook, the 1993 statement also projected a vision of God's universal history with the whole creation, of the human creature as immersed essentially in that history and embedded in nature, and of human caring for nature—not stewardship over nature (this terminology does not appear in the 1993 statement)—as the proper theological framework for interpreting what the human relationship with nature is intended by God to be. The ethic toward nature that the statement recommended was also global and focused on the just claims of the poor and the oppressed around the world. And it affirmed nature's own standing as a participant in God's history with the whole creation.

In retrospect, in my judgment, both the 1972 statement, with its guidebook, and the 1993 statement were paradigmatically shaped by theocosmocentric assumptions. Accordingly, the ethos proposed in both cases was predicated on kinship imagery, rather than on management imagery. And social justice imperatives were remarkably well-integrated into both the substance and the recommendations in both 1972 and 1993.

How much, however, these materials, widely circulated in Lutheran circles as they were, led to corresponding behavioral changes in grassroots Lutheran communities or to organized ecclesial pressure in behalf of substantive theological, liturgical, social, and political change for the better in public arenas is an entirely different matter.

It did appear to me at the time, however (and it still appears to me), that the ELCA church-wide theological impetus that produced those social teaching statements and that first supporting theological document was sustained and developed, along the way, by a number of grassroots theological initiatives in Lutheran circles. I have in mind especially the emergence of the website "The Web of Creation," sparked by the New Testament scholar, David Rhoads, and the Lutherans Restoring Creation movement, also fostered by Rhoads, a venture that championed the establishment of "Green Congregations," "Green Synods," and "Green Seminaries" in the world of American Lutheranism, all in conjunction with a variety of programs and groups focused on ecological issues in Lutheran colleges, universities, and campus ministries.

From this movement, moreover, again at Rhoads' initiative, also arose a proposal for an experimental liturgical lectionary, *The Season of Creation: A Preaching Commentary*, for a portion of the Church Year, focusing on ecological theology and ecojustice issues.[44] Rhoads edited a volume of "Classic Sermons on Saving the Planet," too.[45]

Not to be overlooked, either, has been the marked impact that the social teaching statements have had in the context of Lutheran liturgical life. As a long-standing professional consumer of such services, I often recognized the language and the theology mandated by the social teaching statements in a variety of church publications intended for parish clergy. The prayers made available to congregations through the Lutheran bulletin service in those years, *Celebrate*, regularly included thoughtful references to nature and ecojustice concerns. So did suggested eucharistic prayers in the 2006 hymnal, *Evangelical Lutheran Worship*; furthermore, a number of new hymns reflected creation and ecojustice themes.[46]

The churchwide adoption of those social teaching statements and the implicit sanctioning of the theological guidebook with the first, along with the ensuing churchwide practical initiatives, some of which I have catalogued here, were not a result of some popular fancy, a thought that I have heard from critics over the years, a sort of theological keeping up with the Joneses. Did all this Church reflection and Church activism happen on the basis of what some critics thought of as a general cultural frenzy that came to expression in Earth Day celebrations in those years, which then flowed over into the life of the Church, particularly into the world of the laity, prompting them to call on their denominations to address ecological issues? Was the world, in this sense, setting the mission agenda for the Church? Yes—and no.

The enthusiasm for "environmental issues" generated by the widespread cultural impact of trends collectively known as the Ecology Movement in the U.S. surely prepared the way for the Church-wide actions in 1972 and 1993 and for ensuing practical ecclesial initiatives. But the churchwide support for those statements would not have happened, I am

44. Habel et al., eds., *Season of Creation*. Cf. also the related lectionary aid written from the perspective of the science and theology dialogue, by George L. Murphy, Lavonne Althaus, and Russell Willis: *Cosmic Witness: Commentaries on Science/Technology Themes*.

45. Rhoads, ed. *Earth and World*. For further insightful explorations of the challenge and the promise of ecological preaching, see Schade, *Creation-Crisis Preaching*.

46 Cf., for example, the hymn "Touch the Earth Lightly," hymn number 739, stanzas 1 & 2: "Touch the earth lightly, use the earth gently, nourish the life of the world in our care: gift of great wonder, ours to surrender, trust for the children tomorrow will bear. We who endanger, who create hunger, agents of death for all creatures that live, we who would foster clouds of disaster—God of our planet, forestall and forgive!"

The Theology of Nature as an Emergent Field of Promise

convinced (on the basis of extensive anecdotal evidence), without the practical support of a widespread church ministry that flourished in those years in Lutheran circles (as well as elsewhere): the outdoor ministry movement.

I like to think of that movement as an alternative system of theological education in that era of American Lutheran history. In those years, scores of "Church camps" flourished and served large, even huge, numbers of laity of many ages and many clergy as well, in every corner of the Lutheran Church in the U.S. While there were undoubtedly numerous false starts in those settings theologically speaking (for example a certain overly innocent cozying up to Native American traditions or a facile romanticizing of nature, especially wild nature, in a Thoreauvian mode), serious theological reflection about nature and the global ecojustice crisis was also deeply rooted in many of those settings, sometimes self-directed, sometimes dependent on published works that could willy-nilly be plucked from the works of the Lutheran ecological theologians that I have been discussing.[47]

Prima inter pares, perhaps, among all the Lutheran Outdoor Ministries in this era, from the perspective of ecological theology, was Holden Village in Washington state. A former mining town—which, when it was donated to the Church, brought with it much polluted soil—nestled deep in the wilderness of the Cascade Mountains, Holden Village has fostered serious theological engagement with ecological issues for many years and has, thereby, been a place of renewal for many Christians from all over the U.S., especially those interested in ecological theology. At Holden Village, significantly, theological specialists such as Larry Rasmussen and a wide variety of grass-roots Church leaders, experienced in the struggles for ecojustice, have been teachers, off and on, for many years.

In my judgment, which I cannot document at this point, the often-unheralded work of countless outdoor ministries, like Holden Village, prepared the theological way in Lutheran judicatories in America for the popular churchwide adoption of the aforementioned 1972 and 1993 social teaching statements and for serious-minded follow-up initiatives such as the outreach efforts of Lutherans Restoring Creation.

---

47. Anecdotally, in those years my book *Brother Earth* was regularly read by leaders in outdoor ministry; and I was regularly asked to consult about educational materials in outdoor ministry. I even wrote some of those materials myself, see Santmire, "Introduction to the Theme."

Celebrating Nature by Faith

The First Chapter of American Lutheran Engagement with the Emergent Field of Ecological Theology in Retrospect

At this point, the new theological field we have been considering has indeed shown significant promise in this one historical context.

For many decades since 1961, when Joseph Sittler addressed the World Council of Churches, American Lutheran engagement with ecological theology has proved to be commensurate with the scope of a world in crisis and with the challenge of fostering an ecological reformation of Christianity, because, in varying ways, it has presupposed the theological paradigm of theocosmocentrism. This has been particularly true of the contributions of Joseph Sittler and Larry Rasmussen. But others have played vital roles in this respect, too, as have Church-wide social teaching statements and various practical ministries of the Church. Still, will this beginning have a future? And if so, what should that future be?

One question immediately come to mind. This has to do with a certain pronounced theological diversity in this new field. Each of the theologians whose works I have considered in this and the preceding chapter has moved within the boundaries of ecological theology. But each has also had his own characteristic theological accent, sometimes sharply different from the others: Christology (Sittler), anthropology (Hefner), eschatology (Peters), biblical studies (Fretheim), liturgical studies (Lathrop), historical theological studies (myself), and ethics (Rasmussen). And those engaged in the practical initiatives I have touched on in this chapter have on occasion, if not always, seemed to speak with many tongues, too.

How will those who are now writing or who will hopefully soon be writing the next chapter of American Lutheran engagement with ecological theology respond to what might seem to be, in some respects, that theological glossolalia of the first chapter? Might it be possible to have a more unified Lutheran voice—perhaps even a systematic Lutheran voice—addressing all these issues? On the other hand, perhaps the problem is that the diversity has not been great enough.

This question points to the challenge of hearing other voices, as the new chapter is being written. The figures I have discussed in this chapter, most readers will have noticed, all are men. Yes, they all have been sensitive to a variety of issues that transcend their social locations. But this situation must change. And, as a matter of fact, as the second chapter of Lutheran engagement with ecological theology is now being written, that situation apparently *is* changing, whether fast enough and extensively enough, however, is another question.

Related to this issue is the question of contributions to ecological theology, both academically and practically, from global Lutheran communities. The figures I have discussed all are relatively affluent American scholars. How are American Lutherans to hear other Lutheran voices—not to speak of ecumenical voices and the testimonies of other religious traditions—from the front lines of the global Churches, addressing ecological issues? I remain enchanted, in this respect, with the fruits of Larry Rasmussen's sabbatical adventure of some years ago, when he visited grass-roots, ecologically engaged ecumenical communities in Zimbabwe, Scotland, Alaska, and the Philippines.

Perhaps the single most revealing case for all Lutherans and also others to ponder in this respect is the situation of the Lutheran congregation in Shrishmaref, Alaska. These are Lutherans who belong to an indigenous people which has lived in that area for hundreds of generations. But now their ancestral island home is being washed away by rising waters driven by climate change. Their voices and others like theirs must be heard and given a place in the unfolding next chapter of American Lutheran engagement with ecological theology. That process is already underway.[48] But can it be sustained and expanded?

Finally, moving from the global to the parochial, I offer a personal plea for renewed attention, as the second chapter of American Lutheran engagement with ecological theology is being written, to historical investigations in the theology of nature, both studies of Scripture and postbiblical Christian traditions. We do not want to stumble, due to lack of knowledge, into the pitfalls of the Christian past. Nor do we want to overlook the riches of the Christian past.[49]

---

48. For an entrée into these discussions, cf. the works of Cynthia D. Moe-Lobeda, such as *Healing a Broken World: Globalization and God*; and *Resisting Structural Evil: Love as Ecological-Economic Vocation*. These emergent trends are represented in a number of the essays in Bohmbach and Hannon, eds., *Eco-Lutheranism*.

49. The scholarly study of the theology of nature in the Old and New Testaments has expanded geometrically in recent decades. The trove of these riches, however, is too large to begin to describe here. I have referred to some of such studies in my notes to Chapter 1. Recent studies of historic and current Christian theologies of nature have also been impressive. See especially the two volumes edited by Ernst M. Conradie, *Creation and Salvation*. The single best recent study of historic Christian thought about nature is by Edwards, *Christian Understandings of Creation*. More generally, see the 662-page

Epilogue: From Lutheran Minimalism to Lutheran Maximalism in a Time of Global Emergency

Which brings me at the very end to a single historical issue that must be addressed if the emergent new ecotheological field I have been discussing in this chapter is ever to be widely recognized in ecumenical circles for what it is, and thus be in a position to impact the crisis of our times head-on, as Luther impacted the crisis of his times head-on. This, then, is my question, at the end of these particular explorations. Are Lutheran theologians and practitioners fully equipped today to foment—or to keep fomenting—an ecological reformation of Christianity and to do so with a sense of urgency?

This is how I propose to respond to this question briefly, by raising another. I know that the following question may sound regressive or parochial or even quaint to some, but I believe that it goes to the heart of the matter for those who approach ecological theology as Lutherans or as Lutheran sympathizers in these times. *How do we Lutherans, among others, read Luther?* I have offered a concrete answer to that question in Chapter 2 on Luther's theology of nature. Here I want to speak more generally about what I consider to be the interpretive alternatives.[50]

There are *minimalist* and *maximalist* readings of Luther. The first, in all likelihood, if not inevitably, makes it easy for Lutherans and others to think primarily in terms of the theoanthropocentric paradigm. The second, in all likelihood, if not inevitably, makes it easy for Lutherans and others to think in terms of the theocosmocentric paradigm.

This is Lutheran minimalism. You focus on those theological constructs that drove Luther's reforming zeal at the outset: faith over against works; justification over against sanctification; the theology of the Cross over against the theology of glory; the hearing of faith over against the seeing of speculation; the revealed God in Word and Sacraments over against the God contemplated in nature; the hidden God over against the God who is encountered in, with, and under all things; the Christ who is given for me over against the Christ who is given for the world; the Spirit of God who calls me to faith over against the Spirit of God who hovers creatively

---

reference work, *The Oxford Handbook of Religion and Ecology*, edited by Roger S. Gottlieb, for selected studies of Christian theology, in the context of our global culture. Then I want to mention two studies of individual Christian theologians' approach to nature (Calvin and Tillich), because each one is exemplary in its own way: Schreiner, *Theater of His Glory*; and Drummy, *Being and Earth*.

50. See, further, Santmire, "Healing the Protestant Mind."

over the whole cosmos; and the book of Scripture over against the book of nature.

To that end, you turn again and again to engage the Luther who wrestled with Romans 1:17ff. concerning justification by faith alone, as Luther translated that text. You also elevate the teaching of the Luther who wrote the Heidelberg Disputation in 1518 and who thereby accented the theology of the Cross. *Lutheran minimalism clings passionately to those breakthrough moments in Luther's life and thought,* above all to the reassuring and powerful and liberating Gospel of the forgiveness of sins.

If that is how you read Luther, mainly in terms of his breakthrough moments, then, it seems to me, your own theology and spirituality and discipleship will in all likelihood, if not inevitably, end up being thoroughly shaped by the theoanthropocentric paradigm. Why? Because all those existentially traumatic and spiritually powerful breakthrough moments have to do *with God and with you.* Or, more generically: with God and humanity. Accordingly, you will receive the bread and the wine of the Eucharist as given *for you,* and the theme that the Eucharist is given for the world may fall to the wayside.

And in the spirit of Luther, perhaps one of the greatest polemicists in the history of Christian theology, you may be moved to take your stand against anyone who says, in a manner Kierkegaard would also have abhorred, *both/and.* No, you cannot have it both ways: both faith and works; both Cross and glory; both hearing and seeing; both word/sacrament and creation; both book of Scripture and book of nature; both the Spirit who brings Christ to you and the Spirit who brings Christ to the whole cosmos. You must have it the one Lutheran way, so-called. That is the spirit of Lutheran minimalism in its most contentious form.

On the other hand, would it not be possible to be a Lutheran maximalist? To begin with, you would, of course, *affirm* those Lutheran breakthrough moments and never let them slip through your fingers. But you would also grasp the paradoxical promise of having it both ways: faith and works; justification and sanctification; the theology of the Cross and a theology of eschatological glory; the hearing of faith and the seeing of inspired contemplation; the revealed God in Word and Sacraments and the God encountered in, with, and under the world of nature; the immediately present Christ, given for me, in Word and Sacrament, and the immediately present Christ in all things, given for the whole world; the Spirit of God who calls me to faith and the Spirit of God who hovers creatively over the whole creation; and the book of Scripture and the book of nature.

And you would read and ponder not only Luther's faith-alone commentary on Rom 1:17ff., but also his thoughts about one of his other favorite texts, Eph 4:10, pertaining to Christ: "He who descended is the same one who ascended far above all the heavens, so that he might fill all things" (NRSV).[51] You would, likewise, read not only Luther's Heidelberg Disputation, but also his sacramental reflections about God, in, with, and under all things, in his conversations with Zwingli, and his commentary on John 1, concerning the cosmic Christ, and on Genesis 1 and 2 about God's gracious giving in the whole creation and God's gift of solidarity with the animals to Adam in Genesis 2. You would understand the Gospel not just as the forgiveness of sins, but, in Luther's own words in his Small Catechism, as the forgiveness of sins, life, and salvation. You would receive the bread and wine in the Eucharist as given *for you and for the world.*

Lutheran theological maximalism thus provides us with a way of thinking that can readily be shaped by the theocosmocentric paradigm, which comprehends, as Joseph Sittler first called us to do, all things (*ta panta*). From this angle of vision, everything counts, not just God and humanity, certainly not just God and me. Embracing this theological maximalism will allow us, in turn, even encourage us, I believe, to find new and more forceful ways to address the distress of our earth, as it now groans in travail, and also to celebrate the whole creation, which in its own way is likewise groaning.

Further, Lutheran theological maximalism, shaped, again, by the theocosmocentric paradigm, offers us the impetus we need, I believe, rightly to celebrate the goodness of the whole creation, the miracles of a grain of wheat, the wonders of a child's caress, the glories of coastal and wilderness vistas, and the infinite mysteries of the resplendent heavens above and all around us.[52]

---

51. For an attempt to explicate Luther's "cosmic Christology," which I have highlighted above (Chapter 2, Part 7), see Santmire, "Toward a Cosmic Christology." Also, see the study by the Lutheran theologian/physicist George Murphy: *The Cosmos in Light of the Cross.*

52. This Lutheran maximalism, at home with the theocosmocentric paradigm as it typically is, it appears to me, could also have a sharp public impact, over against the dominant cultural trends of our times, which are so thoroughly shaped by "the information revolution." The presuppositions of that revolution are *radically anthropocentric*, even Gnostic, not cosmocentric.

The information revolution is predicated on the assumption that there is a

This is not to suggest that Lutheran theological maximalism is the whole gospel truth for our times. We Lutherans still have much theological work and much deep soul-searching to do with our two kingdoms ethical heritage, for example

All the more so, time may not be on our side in Lutheran circles, if our *modus operandi* continues to be the often cautious and sometimes self-protecting processes of our scholarly discourse, the measured social teaching statements we produce, after years of study and hearings, and the modest congregational mission initiatives to which many of us have grown accustomed that may accent recycling, say, but not social transformation.

Let us assume, then, that that theological maximalism is what Lutherans—as well as what many other Christians, *mutatis mutandis*—need, in these near-apocalyptic times. The following question still is unescapable: is not the time—the *kairos*—at hand, perhaps as it never has been before, to translate that theological maximalism into public theological praxis and to do that with all the spiritual passion and moral urgency that our times require?[53]

Given the global emergency of the ecojustice crisis that we face today, enormously more pronounced than it was in the nineteen-sixties, when some of us first began to explore what an ecological reformation of Christianity might mean, must not those of us who are Lutherans now reclaim not

---

technological fix wrought by humans that allows humans to overcome any predicament, however global, the laws of nature to the contrary notwithstanding. Nature, in this perspective, is the mere object or the toolbox for the satisfaction of human needs. Nature has no standing, no integrity, no voice of its own.

For an illuminating description of this current culture of power over nature, see Larry Rasmussen, "Next Journey: Sustainability for Six Billion and More," especially pages 102–3.

53. This is not the place to describe the extent and the depth of the global crisis before us. From many voices that might be cited here, consider only one, the well-respected environmentalist David W. Orr, *Down to the Wire*, xii–xiii: "The capacity and apparent willingness of humankind to destabilize the climate conditions that made civilization possible is *the* issue of our time; all others pale by comparison. Beyond some unknown threshold of irreversible and irrevocable changes driven by carbon cycle feedbacks, climate destabilization will lead to a war of all against all, a brutal scramble for food, water, dry land, and safety . . . Sheer survival will outweigh every other consideration of decency, order, and mutual sympathy. Climate destabilization will amplify other problems caused by population growth, global poverty, the spread of weapons of mass destruction, and the potential impact of high consequence events that have long-term global consequences . . ." I also highly recommend the works of humanist/environmentalist/activist Bill McKibben, beginning with his classic, *The End of Nature*.

only the full breadth of Luther's theology, as Lutheran maximalists, but also *the apocalyptic sensibility* which drove Luther's own reforming zeal?

Luther was willing to confront pope and emperor in the name of the gospel truth. He was willing to sacrifice all that had hitherto been of existential import to him in the furtherance of that cause. Is not the time at hand for those of us who treasure Lutheran theology, now in its maximalist expressions, to confront the principalities and powers of our own world with the same kind of apocalyptic intensity? Likewise, for members of all other Christian communions. Has this not become for all who aspire to be followers of Jesus a time of passionate public witness and resolute communal action, a *status confessionis*, as Luther's time was for him?[54]

---

54. I am aware that the expression *status confessionis* is ambiguous, both historically and theologically. For a thoughtful exploration of this ambiguity, see TeSelle, "How Do We Recognize a *Status Confessionis*?" The expression is rooted deeply in Lutheran history, from the mid-sixteenth century in Germany, through the German Church struggle of 1933, to the 1977 declaration of the Lutheran World Federation that apartheid is a heresy. Minimally, the idea is this, in Teselle's words, that "to declare a *status confessionis* is to say that the time has run out, that toleration has reached its limits, that a line must be drawn. It is to say that the time is 'an evil time' (Amos 5:13), but one in which we may no longer keep a prudent silence" (78). Given the severity of the global crisis (see note 53 above), does the Church today have any other option than, in the name of God's love for the whole creation, especially for the downtrodden of the earth, to publicly and zealously speak the truth to power and to its own members, with a willingness to put its own body at risk? Isn't the burden of proof on the shoulders of those Christians, particularly Lutherans, who would maintain that ours is *not* a time of *status confessionis*?

5

Celebrating Nature by Faith
A First-Person Theological Narrative[1]

AN EXPERIMENTAL FILM I once watched began with this improbable image: a bicycle wheel splashing through a puddle of water on a city street. A more conventional film might well have begun with the camera scanning the whole city—of New York, it turned out—from above. You would have seen the skyscrapers of Manhattan from a distance, perhaps, then a busy city street down below, full of traffic and with crowded sidewalks. Only then, as the camera zoomed down, would you have been presented with the image of a bicycle rider and, finally, the wheel of that bicycle splashing through that puddle of water.

Each approach has its own kind of promise, whether one begins with the overview or the particular image. Thus far, in these five studies in Reformation theology in this, our era of global emergency, I have been most concerned with the overview. I first explored what the Scriptures tell us about living with nature, accenting and explicating the theology of partnership as an alternative to the widely championed theology of stewardship (Chapter 1). I then offered a study of the works of a venerable theologian from the relatively distant past, who still speaks to many of us and whose theology of nature, were it better known, could be an inspiration to many, Martin Luther (Chapter 2), and the witness of a leading theologian from our own era, Joseph Sittler (Chapter 3), whose voice needs to be heard in these times more than ever. But Sittler by no means stood alone in his own

---

1. These reflections are a much expanded and revised version of the essay I published in Nelson et al., eds. *Theologians in Their Own Words*, ch. 18.

time as an ecological theologian writing in the tradition of the Reformation in the U.S. He was joined by a number of confreres, all of whom were concerned both with ecological and justice issues but who wrote from a variety of perspectives (Chapter 4). In this context, I also reviewed a number of related, theologically inspired practical initiatives that were critically influential in their own ways.

At the end now, I turn, figuratively, from the overarching views of a Manhattan seen from above, to the sight of a bicycle wheel splashing along a Manhattan street. My purpose at this concluding moment is not to draw any conclusions, but to particularize the whole discussion in terms of my own experience. My hope is that these autobiographical reflections at the end will reveal some of the existential dynamics that underlay the more wide-ranging discussions in all the previous chapters and, to that degree, make those explorations more accessible and therefore more understandable, if not more compelling, in retrospect.[2]

The Making of an Ecological Theologian

Although I did not know it at the time, my life as an ecological theologian had already begun during my college years, when I encountered the personae and the teachings of two theological giants, both of whom had been nurtured by the traditions of Luther in Germany: Paul Tillich and Dietrich Bonhoeffer. My budding life as an ecological theologian was also enriched a decade later, again not consciously and perhaps improbably, by two American figures who were not theologians, at least in the usual sense of the word, Henry David Thoreau and John Muir.

Tillich had just arrived at Harvard when I was a sophomore. He immediately began teaching his famous Theology of Culture class for undergraduates. As I sat there in rapt attention week after week, understanding some of what he said, I fell in love with Tillich's approach to theological discourse, a love which I have never lost. Subsequently I enrolled in every course he ever offered at Harvard, first as a college student and thereafter when I was doing graduate work at Harvard Divinity School.

---

2. These autobiographical reflections highlight the trajectory of my life as a professional ecotheologian and a practicing pastor. Other dimensions of my personal history will, as a matter of course, be by-passed, among them my life as a marriage partner. For the latter, see my essay, "Images of an Ordinary Conjugal Spirituality."

Celebrating Nature by Faith

I am not sure when I read Tillich's stellar essay "Nature and Sacrament" for the first time, but I do remember that it touched me deeply.[3] I had been reared in what I later came to think of as sacramental Lutheranism, with much attention, literally, to the earth. My father and mother were not only faithful participants in our congregation's sacramental life, they also engaged themselves and their children in the gardening life, caring for flowers and shrubs and vegetables and specimen trees. During World War II, the whole family worked on a huge "victory garden," as many American families did in those days. I still remember the joys of planting and cultivating and even weeding. That was a heroic thing for a young boy to do, in those days. After the war, my parents introduced my brother and me, and my then new sister as well, to the vistas of many of the U.S.'s great national parks. In our own modest ways, ours was a family that was rooted in nature, at hand and afar. Hence Tillich's teachings about the sacramental values of nature and about the awesome ambiguities of nature immediately resonated with me when I first encountered them in my undergraduate years. I had discovered in Tillich's lectures, without knowing it, what was to become a driving theme of my whole vocational life.

It was no accident, then, that, upon entering the Divinity School, I found a way to devote much of my extracurricular reading time to the works of two great American champions of nature, the aforementioned Henry David Thoreau and John Muir. I cannot remember now how that passion for those two prophets of the wilderness first developed. Perhaps it was when I was introduced to them by the then much revered historian of American culture, Perry Miller, some of whose lectures I audited during my college years.

Of the two great American prophets, I remember identifying most with Muir. I think that this was because Muir himself had been reared according to the nature-loving canons of John Calvin's theology. Muir was deeply conversant, too, with King James English and the world of biblical metaphors, in particular (his father made him memorize the whole New Testament as a child). Muir's descriptions of vast glaciers or brilliant poppy fields or grand tillandsia-draped oak trees were typically shaped by biblical imagery. I especially remember one of his descriptions of an Alaskan glacier, which, he said, "came forth in the fullness of time."

Thoreau was something of a puzzle to me and still is in some ways, although the raw spirit of his writings, especially on Cape Cod and the Maine

---

3. Tillich, "Nature and Sacrament."

woods, spoke to me back then and still does. Curiously, even though I had lived in the Boston area, off and on, for many years, only after my retirement did I find an occasion to visit Thoreau's cabin and Walden Pond. But, when I did, and when I then reached down to put my hand in the Walden waters so that I could make the sign of the Cross over my heart, it was as if I had been there many times. That experience was a kind of homecoming.

The Encounter with German History

And Bonhoeffer? My encounter with this heir of Luther was much more urgent and in many ways much darker than my engagement with Tillich and prophets like Muir and Thoreau. I grew up in a well-protected cultural environment in Buffalo, New York in the nineteen-forties, proud of my family's German heritage. But then there was the Holocaust. For a long time, I knew nothing about it. Although I was a regular Church-goer in my teens and a participant in numerous Church youth events in those days, I had never even heard the word Holocaust during my high school years, as far as I can remember.

Toward the end of the war, I did see what must have been some only slightly edited newsreel films about the liberation of some of the camps. The images of the piles of bodies and the emaciated survivors horrified me. But I must have repressed those experiences. As I recall, I never talked about them with anyone back then. Nor was I ever called upon, as I remember, at home or in school or at my Church, to discuss what I had seen in those films. I am not sure when the very expression, the Holocaust, gained currency in my social and religious world. Remarkably, I remained consciously oblivious to the Holocaust through the end of my high school years in 1953.

All this changed one day, as I sat in a German history class when I was a college sophomore. The instructor began to review the story of the Holocaust and then to narrate how many "good Germans" turned the other way when Jews were carted off to the camps. He talked, too, about Lutheranism's long tradition of hostility to the Jews in Germany, beginning in a most ugly fashion with Luther himself. The instructor also described the Lutheran tradition of unquestioning obedience to the state, predicated on a reading of Romans 13:1 (KJV): "The powers that be are ordained by God." That faith debilitated the consciences of many German Lutherans in those days, said the instructor.

But those were *my* people! That was the Church tradition which had so powerfully nourished me since I had been a child! Not a few Lutherans had become "German Christians," said the instructor. Me? World War II had long been over, but I had never even thought about the meaning of the Holocaust! Why hadn't we discussed it at my home Church? Why hadn't synodical bodies addressed this issue or speakers at youth rallies or at Bible camps? Maybe they did. But if so, I had never really listened. In that college class that day, I found myself at the edge of tears. I still feel those tears.

Back then, I did not know what to do with my shockingly late discovery of the Holocaust. At the time I was living with four Jewish roommates. We talked about everything, but not the Holocaust in any depth, as far as I can remember. By default, if not by intention, I kept all my Holocaust feelings of dread to myself, for most of the four years that the five of us lived together. During this time, I did go to see my own pastor at University Lutheran Church and began to share my grief and my guilt with him. At one point, he asked me whether I knew the story of Dietrich Bonhoeffer. I did not. But I soon did, and more.

I was able to do my undergraduate honors thesis on the German resistance to Hitler, and Bonhoeffer was one of the figures I studied in some detail. In the course of my research, I also encountered the history of the Confessing Church in Germany, with which Bonhoeffer had been associated, along with a few other leading theologians of the day, such as Karl Barth. While the established churches, both Catholic and Lutheran, had either found ways to support National Socialism actively or to step aside and let the Nazi project run its own course, members of the Confessing Church took a public stand against the regime, not always effectively, but sometimes dramatically and sometimes with fatal results.

In the course of my studies, I also worked through materials about the Nazi approach to—nature! I discovered that the Nazis had championed nature in their own demonic way, a term that I had, by then, learned from Tillich. As I have already noted in this book, the Nazis championed heroic violence, an ethic of "blood and soil," *Blut und Boden*, akin to what they thought to be the dominating animal violence of the primeval German forests. And here *I* was, an American nature-lover of German descent, who had managed to live through the war years and beyond, in almost total ignorance of the Holocaust!

Later, during my Divinity School years, as I reflected about my own love for nature, highly ambiguous as I had come to realize that it had been,

I began to develop a more nuanced view of the witness of writers like Thoreau and Muir. Although Thoreau, for example, was disgusted by slavery and by American participation in the war against Mexico and although, as a result, he took a public stand against his own government, which he described in his essay on civil disobedience, Thoreau constantly mounted a critique against what he perceived to be the evils of urban life. In contrast, he championed what he took to be the vital and pure ethos of the American wilderness. His ethic, in that context, was essentially escapist. Likewise for Muir. Following Thoreau, Muir regularly rejected what he considered to be the dirt and the superficiality and the money-grubbing mores of the city, in favor of what he thought of as the cleansing balm of the wilderness.

The Emergence of a Theme: Ecology and Justice

In retrospect, I can see that my undergraduate and early graduate studies bequeathed to me two of the themes that have been at the forefront of my theological mind and heart ever since, the theology of nature—or ecology—*and* the theology of justice. In light of the National Socialist exaltation of nature, never again would I be able, if indeed I had ever been, simply to be a nature romantic.

Nature, I had learned (although I did not know this language at that time), was in significant measure a social construction, for better or for worse. In light of the fact that the scales fell from my eyes when I looked at my own treasured Lutheran tradition, moreover, I would never again be able to think of the Church, or anything else for that matter, without also thinking of the claims of justice for the despised and the oppressed.

Give me the sometimes flawed socio-political commitments of a Bonhoeffer or the sometimes ambiguous commitments of a Confessing Church any time, I thought to myself. Even Tillich's commitment to Religious Socialism, politically ineffective as that movement turned out to have been, at least had the merit of dealing head-on with the social and political issues of his day, thus steering around any kind of romantic escape into the embrace of nature, as numerous intellectuals and artists in Weimar Germany, Tillich's spiritual home at that time, had already decided to do.

This commitment to both of these themes, ecology and justice—I highlighted them sharply in a 1976 article in the *Christian Century*[4]—it

---

4. Reprinted in MacKinnon and McIntyre, eds., *Readings in Ecology and Feminist Theology*, 196–207 (ch. 5).

turned out, left me in what I perceived to be a remarkably lonely position, especially in the early years of my theological career. In that era, both within and beyond the Church, numerous advocates spoke up publicly in behalf of defending nature. The American scene did not lack people who championed the cause of "the land" or "the wilderness," those venerable themes of American cultural history. Numerous advocates, again, both within and beyond the Church, also dedicated themselves to fighting for justice, as the struggles against racism and the war in Vietnam moved more and more to the front page. But in those days few were inclined to speak and act in behalf of *both* nature *and* justice.

Often, therefore, I found myself as a young theological student, talking justice with the ecology people and ecology with the justice people. This has been a struggle for me ever since. There has indeed been a deep-seated tension between the concerns of the ecology party and the justice party in America, a tension, indeed, that cannot easily be resolved. But I am getting ahead of myself.

Encountering Karl Barth and Theoanthropocentrism

It was time for me to write my doctoral dissertation. Of course, for me, given my own intellectual and spiritual history, my thesis *had* to deal with the theology of nature. But with what focus? With what argument? As I thought about it, a thesis on nature would have to move against the grain of the biblical and theological projects that were well established at that time in Reformation and Catholic circles.

In my formative theological years in the late nineteen-fifties and early nineteen-sixties, as I already have had occasion to observe, biblical theology was dominated by a heady and self-conscious theoanthropocentrism, the "God-who-acts" theology of G. Ernest Wright (with whom I took two courses), who pitted what he thought of as historicized biblical faith over against what Wright also called the nature religion of the Canaanites. New Testament studies was then to a significant degree under the sway of the demythologizing program and the existential analysis of Rudolf Bultmann, then a favorite of college chaplains and university religion teachers. Bultmann argued, revealingly, that the groaning of the whole creation that Paul talked about in Romans 8 referred to the groaning of the *human* creation only! Like his neo-Kantian mentors and Kant himself, Bultmann in effect handed nature, which he tended to view in mechanistic terms, over to the natural scientists and the entrepreneurs.

In systematic theology in those years, likewise, the theoanthropocentric theology of "salvation history" (*Heilsgeschichte*) was much in vogue. Thus, as I already have had occasion to observe, the famous neoorthodox theologian and sometime theological opponent of Karl Barth, Emil Brunner, once stated that nature is merely "the stage" on which God's history with humanity unfolds.[5] I was told, indeed, as I have likewise already noted, a few years later by the American Kantian theologian who was to become my dissertation advisor, Gordon Kaufman, that "theologians need not concern themselves with nature."

That I had learned important things about nature in the course of my studies, above all from Luther and from Tillich, is another story. That I began to see that there was another, non-anthropocentric way to approach the Scriptures, especially when I read what was to become an epochal 1963 article by the Lutheran scholar-pastor, who had already become a mentor for me, Krister Stendahl, "The Apostle Paul and the Introspective Conscience of the West," is also another story.[6] Still, if I were to write a dissertation on the theology of nature, I concluded at that time, I would have to more or less go it alone as far as biblical research was concerned, given the theological world in which I was living.

Nor was there much theological discussion in the late fifties and early sixties, even less public awareness, of what was already in those years becoming a major challenge for our species, the environmental crisis. This widespread environmental indifference only began to give way somewhat in the wake of Rachel Carson's *Silent Spring* in 1962 and Stewart Udall's *The Quiet Crisis*, in 1963.[7] I myself had no idea in those years that signs of the crisis were then already apparent. I did not read the Carson and Udall books until the late sixties. I was too busy with my dissertation and a year of study in Germany and travel south of the Alps.

The dissertation topic I had chosen? Karl Barth's theology of nature. I could have opted to work on Tillich, whose writings I knew well by then (I was a member of what I think was Tillich's last graduate colloquium at Harvard, which focused on his *Systematic Theology*). But my theological passion in those days—and to this very day—was not first with a Tillichian "apologetic theology" which followed a "method of correlation," however

---

5. Brunner, *Revelation and Reason*, 33n.

6. Stendahl, "Apostle Paul and the Introspective Conscience of the West," in Stendahl's volume, *Paul among Jews and Gentiles*; first published in the *Harvard Theological Review*.

7. Carson, *Silent Spring*; and Udall, *Quiet Crisis*.

important that was and is, it was first and foremost with the theology of the Word of God. Tillich had made a place for what he called "kerygmatic theology"—the term widely used in Church circles in those days to describe theologies of "the message," that is, the Word of God—and I found myself eagerly exploring that place.

This was no doubt because of my existential engagement with the story of the Confessing Church and with the figure of Bonhoeffer, in particular. For such reasons, I gravitated as a matter of course toward a dissertation on Karl Barth, who had been deeply involved in the Confessing Church himself.

Along the way, I also thoroughly immersed myself in Luther studies, under the tutelage of the then soon-to-be renowned Dutch scholar, Heiko Oberman. I had worked on Luther, too, during a year's study at the Lutheran Theological Seminary at Philadelphia with that institution's systematic theologian at the time, Martin Heinecken. He interpreted Luther both historically and existentially, and was particularly helpful to me in opening up Luther's rich theology of nature. Those Luther studies formed a kind of bridge for me to Barth, who himself constantly engaged the theology of Luther (especially in Barth's long, historical footnotes in his *Church Dogmatics*). So it was the theology of the Word of God, rather than the theology of correlation, that most claimed my mind and heart in those days—and still does.

But I chose to come at Barth, brashly perhaps, not on his own terms, but on my own. I decided that I wanted to study his theology of nature, notwithstanding the fact that Barth had announced in volume three of his *Church Dogmatics,* as I mentioned at the outset, that there is *no such thing* as a legitimate theology of nature. It is easy to understand why Barth would have been prompted to take that position, given his opposition to the Nazis and to their—demonic—theology of nature.

Barth self-consciously defined theology as theoanthropology—a choice on Barth's part that I have noted often in this book—as a doctrine of "God and man" (the patriarchal language, sadly, that we all used in those days). I argued that by thus focusing his theology on God and humanity, Barth had baptized a non-theological doctrine of nature by default, the instrumental, utilitarian, and mechanistic view championed by bourgeois society in the West. Better, I maintained, self-consciously develop a theology of nature on biblical grounds (conversant with the findings of modern science) than to end up in that kind of theological dead-end.

Although I did not deal with Barth's sacramental theology in my dissertation, that theme was very much on my mind, too. How could Barth ever envision a theology of the real presence of Jesus Christ in Word and Sacraments, as Luther had done, I wondered, if nature, the world of material existence, was only something to be viewed as a symbol or a resource or an instrument or even as a machine?

In this respect, my thinking had been shaped by a careful re-reading of Tillich's seminal essay, "Nature and Sacrament," to which I have already referred. There Tillich showed that the modern Protestant theology of nature was an expression of the spirit of the victorious bourgeoisie. Hence I asked: was *that* very spirit lurking deep within the monumental argument of Barth's *Church Dogmatics*? I concluded that it was, notwithstanding Barth's own laudable, but finally, to me, unconvincing efforts in the *Church Dogmatics* to affirm the goodness of the whole creation.

The Theological Life and the Challenges of Discipleship

No sooner was my dissertation completed than I found myself unavoidably thrust into the—sometimes poignant, sometimes silly, often passionately committed—era of the late nineteen-sixties and the early nineteen-seventies as a college chaplain, first for three years at Harvard, based at University Lutheran Church, then for thirteen years at nearby Wellesley College. I did a lot of theology, as I thought of it in those days, on my feet.

In that era of my life, I was a mostly low-profile church activist and theological essayist, and I was occupied with my liturgical and pastoral duties. I stood with students who sat in on the Mallinckrodt Building at Harvard, protesting the role of Dow Chemical in the Vietnam War. I served on the steering committee of what was to become a national movement, Vietnam Summer. I walked, from time to time, with a Lutheran pastoral colleague, the Rev. Vernon Carter, an African-American community leader, who picketed the Boston School Committee for 114 days, demanding an end to segregated schools (the then famous, now infamous, Boston school busing court decision arose in that context). I became a wary witness in an inner-city police-station once, when the call had gone out to clergy in the area to be visibly present in such places during Boston's "civil disturbances," in order to help guard against police brutality. I stood in silent vigils in a suburban town square many times, gave numerous speeches against racism, against the war, and in behalf of the environment, all with the theology of the Confessing Church in the back of my mind.

Celebrating Nature by Faith

My learnings about racism, particularly about what is now instructively called white privilege, came slowly. Growing up, I don't recall ever knowing any blacks face to face, except for—not surprisingly—my family's weekly cleaning lady, Emma. My first real encounter with African-American solidarity came during one summer when I was in college. I worked the night shift on the ore docks at Bethlehem Steel, just outside of Buffalo. I was the only white in a crew of twelve. We rarely spoke.

One night, shoveling ore, deep in the dimly lighted cavernous hold of a great lake freighter, trying to stay out of the way of the colossal scoop from above that was roaming around down there swallowing up the ore, sixteen tons at a time, one of the crew members, whose name I never learned, saved my life. He pushed me out of the way of the scoop when it was swinging right at me. Without that intervention, that scoop would have crushed me dead against the steel-plated wall next to which I had been shoveling. Many years later, I came to appreciate that Christ-like act much more deeply than I had at the time.

I entered into a different kind of world when I was serving at Wellesley College. I got to know several of the then dozen or so black students (among a student body of some twelve-hundred) pastorally, which was a revelation for me of a more intimate kind. This is one story I was told back then by a young black woman, in tears: a white student got on to the dormitory elevator with her; neither knew the other; the white student asked the black student, "What is it like to be a Negro?"

It turned out that those years at Wellesley College were a good era to be interested both in ecology and justice, theologically as well as pastorally, notwithstanding the inherent difficulties in holding those two concerns together. People at that time, for better or for worse, looked to clergy for prophetic leadership. At Wellesley, I worked on a number of justice issues—like recruiting more black students for the College—with a student leader, Hillary Rodham, who had a future of some fame before her; but in those days she was only one among many likeminded student activists, who were morally driven by the issues peace and justice.

I wish I had kept copies of my commencement prayers from those years, which became a kind of cause celebre at the College, for reasons that I never fully understood back then. In retrospect, it appears to me that those prayers were striking to many in those academic audiences, because the prayers were expressions of a kind of public theology that circumstances had been calling many of us to develop. *Mirablile dictu*, they were, as a matter of course, petitions devotionally addressed to *God* and not just pious and politically correct nostrums, like all too many commencement prayers at that time, utterances which a friend of mine once called vegetable piety ("lettuce affirm our hope and lettuce minister to the poor").

In the late sixties and the early seventies, also, I also found it relatively easy to publish op-ed pieces and other essays on the themes of ecology and justice, in journals like *Dialog* and in the local press. During those years, too, the North Carolina biologist, Paul Lutz, and I co-authored a book intended for popular consumption, *Ecological Renewal*.[8] Presiding over the Sunday Chapel services at Wellesley College, and its occasional public lectures, also gave me opportunity to provide a platform for and to meet and converse with leading lights of "the Counter-Culture" of the late sixties and early seventies, renowned figures such as William Sloane Coffin, Andrew Young, Jesse Jackson, James Cone, Jane Fonda, Katie Day, and progressive Catholic priests like Anthony Mullaney and James Carroll.

In those years, I discovered, too, what was then for me a new wave of theological feminism. I brought theologians Mary Daly and Rosemary Radford Ruether to the College Chapel, for example, and became a public advocate of their right to a hearing, in a setting where, in those days, both faculty and students were not always interested in feminist thinking, or even were hostile to it. I also began to develop what was to become a deep interest in theological feminism at that time, too, not chiefly by any prescience, I hasten to add, but mainly out of a sense of professional self-preservation. As a male chaplain and teacher at an all-women's college, I *knew* I had to be out in front of the curve on that one.

I was particularly taken at that time with the developing thought of Rosemary Radford Ruether and with the emergence of ecofeminism as a central theme in her work. Along the way, I taught what turned out to be the College's first course in ecofeminism. Although, moreover, I have never written as an ecofeminist (for good reasons, I think), I have always consciously tried to ask the questions raised by thinkers like Ruether and, later, Sallie McFague and Elizabeth Johnson, in order to be as sure as possible that I was "for them," however implicitly, rather than "against them." Likewise for other theologies of liberation, which I read avidly in those years, above all the works of James Cone.

Such interests—the Confessing Church, the theology of liberation—have stayed with me ever since. They came to their most visible expression for me some years later, first, in an essay, "The Liberation of Nature: Lynn

---

8. Lutz and Santmire, *Ecological Renewal*.

White's Challenge Anew,"[9] in which I argued that the liberation of nature goes hand in hand with other forms of liberation, then in my theological memoir, *South African Testament: From Personal Encounter to Theological Challenge*,[10] a short book based on a firsthand engagement with the South African Church and with the apartheid system at the height of its power.

Engaging the Emergent Global Ecological Crisis Theologically

It was while I was at Wellesley that I produced my first book, *Brother Earth: Nature, God, and Ecology in a Time of Crisis*.[11] This 1970 work was later called a "neo-Reformation" study by Claude Y. Stewart, Jr. in his book, *Nature in Grace: A Study in the Theology of Nature*[12] in which he compared my modest efforts with those of two theological giants, John Cobb and Pierre Teilhard de Chardin. I was happy to be so characterized, since the two theologians whose works most shaped the argument of *Brother Earth* were Luther and Calvin.

Over against what I then called the "exclusive theoanthropology" of Barth, whereby the main objects of theological reflection are God and humanity, as I have oft noted in this book, I proposed that both Luther and Calvin, notwithstanding their own commitments to a certain kind of theoanthropocentrism, in fact thought more often than not in terms of an "*inclusive* theoanthropology." Their theological framework was tripolar, not dipolar. As a matter of course, they took God, humanity, *and* nature seriously. More particularly, I explored Calvin's rich theology of Divine providence in nature and Luther's profound conceptualization of God as "in, with, and under" the world of nature.

I also sought to undergird my argument in *Brother Earth*, necessarily so, given my commitment to a Word of God theology, by developing my own biblical interpretion. I was encouraged in these efforts by Stendahl's non-anthropocentric reading of Paul, particularly by Stendahl's accent on the world-historical, even cosmic, meanings of Paul's treatment of the controversies over Jews and Gentiles in the early Church. I continued the practice of doing much of my own biblical exegesis for many years, with

9. Santmire, "Liberation of Nature."
10. Santmire, *South African Testament*.
11. Santmire, *Brother Earth*.
12. Stewart, *Nature in Grace*.

fear and trembling. But I had no other choice, I felt, since most biblical scholars in those days, as I have already observed, either were not interested in the theology of nature or they uncritically assumed that the Scriptures are fundamentally anthropocentric.

Brother Earth also employed a kind of correlative method, undoubtedly a sign of Tillich's influence on my understanding of the task of theology. Following the lead of historian Perry Miller, to whom I have already alluded, I diagnosed a schizophrenia in American culture between nature and civilization.[13] In America, I suggested, the ancient dichotomy between the country and the city had become a kind of socio-political obsession. Thus American culture did indeed have a historic fascination with the wilderness, contrasted to the alleged impurities of the city—a theme articulated classically by Thoreau. But American culture simultaneously championed "manifest destiny" and "progress"—witness Emerson's celebration of the railroad—and as a result, typically showed little regard for wilderness values.

Christian faith offers an answer to that schizophrenia, the split between nature and civilization, I argued. That answer, I maintained, is evident above all in the testimonies of biblical prophets like Deutero-Isaiah, but also in traditional Christian thinking about the Kingdom of God as the end (*telos*) of all things. Nature and civilization are kin, I maintained, insofar as both are rooted in and shaped by God's immanent providence and by the coming Kingdom of God.

Brother Earth had a number of liabilities, I now realize. The title itself was problematic. It rightly suggested kinship as the normative human relationship with nature; but I resisted then (for some good reasons, but mostly for bad reasons) thinking of nature metaphorically as female. I also all too easily affirmed a Kingdom of God theology, blithely unaware of the then emerging feminist critique of such symbols. And, not unrelated, I was unaware that my theology of human dominion, at points in that book, did not sufficiently guard against the inroads of modern Protestant/Capitalist/Marxist understandings of human dominion over nature as domination. Still, I think that that book did make at least two significant contributions to the then emerging American discussions of the theology of nature.

First, I insisted on the theme that God has a history of God's own with the vast world of nature, apart from nature's meaning for us humans (by 1970 I was regularly talking about "the integrity of nature"). Second, I drew

13. See, e.g., Miller, *Errand into the Wilderness*.

on the argument of one of my first published articles to which I have already referred, "I–Thou, I–It, and I–Ens,"[14] which was a conversation with Martin Buber, to identify a human relationship with nature and with God in nature that did not turn nature into an "It." Analogous to an I–Thou relationship, I maintained, an I–Ens relationship with a tree, was not objectifying, but neither was it strictly personal (humans do not converse with trees). More particularly, I envisioned the human-nature relationship in terms of wonder and respectful reciprocity. All this I set forth, as a self-conscious, American neo-Reformation thinker, in conversation with Luther and Calvin—and Muir. I am grateful that, beginning with Mary Daly's first edition of *Beyond God the Father*[15] that conceptuality of an I–Ens relation has found a place in some works by ecological thinkers and even in some works by mainline theologians.

Brother Earth also left me with an unfinished theological agenda. The keystone of the argument of the book was its christological center. But my description of that center was underdeveloped. Perhaps that was because I had yet to come to terms, in one way or another, with Barth's famous "christological concentration," which was much under discussion in theological circles in those days. Be that as it may, I began to think about such matters more and more in ensuing years, particularly as I came under the influence of Joseph Sittler during the nineteen-seventies. I had of course read Sittler's famous 1961 address to the World Council of Churches in New Delhi, which I highlighted above, where Sittler called for a new "cosmic Christology," but in those days I was preoccupied with other things like walking picket lines.

That lack of attention to Christology changed as I developed a personal relationship with Sittler, during the time when he and I were the theologians selected to help write the 1972 statement and theological study guide on the environment for the Lutheran Church in America. He graciously befriended me and publicly affirmed my work. In turn, I sat at his feet and particularly benefitted from reading his *Essays in Nature and Grace*.[16] The theme of developing a cosmic Christology has preoccupied me and challenged me to this very day.[17]

14. Santmire, "I–Thou, I–It, and I–Ens."
15. Daly, *Beyond God the Father*.
16. Sittler, *Essays in Nature and Grace*.
17. See Santmire, "Toward a Christology of Nature"; and Santmire, "So That He Might Fill All Things." I also have addressed this theme in Santmire, *Before Nature*, 157–69, along with similar explorations in this book, in the context of my discussions of

Celebrating Nature by Faith
Exploring the Ambiguous Ecological Promise of the Christian Tradition

But, in the late seventies, I bracketed such constructive challenges, in favor of a historical task: revisiting the classical Christian theological tradition from the perspective of ecological theology. Ironically, perhaps, I only became aware of what James Nash called "the ecological complaint against Christianity" relatively late in the day. My interests in the theology of nature had been well-established by the time I first read the now ubiquitously cited 1967 article, "The Historical Roots of Our Ecologic Crisis," by the historian, Lynn White Jr., to which I have already referred more than once in this book.

In that article, White charged that historic Christianity must bear a "huge burden of guilt" for the environmental crisis.[18] That enormously popular expression of the ecological complaint against Christianity, along with my growing awareness of the severity of the environmental crisis itself, gave me a new sense of urgency about the theological path on which I had already embarked. As a matter of course, then, I referred to the White thesis in the Preface to *Brother Earth* and I offered that book, in part, as an answer to White's contention that Christianity has, with rare exceptions (such as the witness of St. Francis), been ecologically bankrupt.

During the seventies, White's argument had become the mantra of many academic critics of Christianity and even of some theologians, among them Christian feminists and advocates of Native American spirituality. Of particular importance to me, Gordon Kaufman, who had supervised my dissertation on Barth and who, as he told me a few years before he died, began to shift his own thought about the theology of nature in response to my study of Barth, publicly launched what was for him a new theological program in 1972. Kaufman came to believe, for his own contextual reasons, that, in effect, Lynn White Jr. was right, that historic Christianity *was* bankrupt ecologically and that it therefore must be totally reconstructed.

As the 1970s progressed, I encountered even more of this kind of thinking on college campuses and, surprisingly, in some church circles— particularly in a few outdoor ministries of the church. I myself had been working all along with a quite different reading of the classical Christian tradition, so, in my available scholarly time, I began to devote myself to

---

Luther (Chapter 2) and Sittler (Chapter 3).
18. White, "The Historical Roots of our Ecologic Crisis," 1206.

developing a fresh interpretation of classical Christian thought about nature. It was a long gestation period. There was much work to be done.

The eventual result was my study, *The Travail of Nature: the Ambiguous Ecological Promise of Christian Theology*,[19] thankfully endorsed by John Cobb, Langdon Gilkey, and, *mirabile dictu*, Lynn White, Jr. himself. I originally intended *Travail* to be a shot across the bow, as it were, a kind of public theological announcement that indeed there are hidden ecological riches in classical Christian thought, notwithstanding a whole range of sometimes profound ambiguities regarding the theology of nature.

I approached the subject archeologically, as I said in the book itself, only sinking down a few trenches into the tradition, so to speak, to see what I might find as a way to encourage others to begin to excavate the whole site. I singled out several theologians for special attention: Irenaeus and Origen; the young and the mature Augustine; Thomas, Bonaventure, and Francis; Luther and Calvin; Barth and Teilhard de Chardin. Invoking a method of metaphorical analysis, I identified two major Christian ways of thinking about nature throughout the ages, the one ecological and theocentric, the other spiritualizing and anthropocentric.

Soberingly, the wave of historical studies of classical Christian thought about nature that I had hoped would appear in subsequent years after the publication of *The Travail of Nature* did not quickly begin to emerge. Thankfully, though, that situation has begun to change, sometimes with very promising results.[20]

Advocating for a Theology of Nature

With the historical work I had envisioned completed in *The Travail of Nature,* I still had the feeling—sometimes forcefully, as I became a kind of circuit-rider, speaking to a variety of Church and academic audiences around the country—that the Church, in its various formations, still very much needed to claim the message about the ecological promise of Christian theology as quickly as possible. Driven by that feeling, I decided that it might be helpful for me to pull together a variety of theological themes in book form in order to speak directly to the American Church's clergy and lay leaders with a summary of how Christians can think about the issues of theology, ecology, and justice in our times, drawing, wherever possible,

---

19. Santmire, *Travail of Nature*.
20. See the extensive literature cited above, Chapter 4, n. 49.

on the theological riches of the past, both biblical and traditional. Hence the publication of my volume *Nature Reborn: The Ecological and Cosmic Promise of Christian Theology*.[21]

This popularizing book might still serve as an introduction to ecological theology. In it, I describe what I think are the theological options today: the way of the *reconstructionists* (Christianity is ecologically bankrupt; let us begin anew), the *apologists* (Christianity has all it needs in its doctrine of stewardship of creation; let us interpret it), and the *revisionists*, in whose ranks I number myself (Christianity has had an ambiguous ecological history; let us reclaim its ecological riches wherever we can).

Pursuant to my own revisionist agenda, in *Nature Reborn* I took issue with those who uncritically accepted the ecological complaint against Christianity (with particular reference to Matthew Fox). I called attention to neglected ecological themes in biblical theology while at the same time I argued that some major expressions of Christian theology, however relevant they might seem in this era of global crisis, are ecologically suspect (citing the exemplary case of Teilhard de Chardin for his spiritualizing tendencies). I concluded with brief discussions of the ecological dimensions of Christian ritual, spirituality, and ethics.

Nature Reborn was a kind of comprehensive statement of what used to be my standard "stump speech," as I traveled around the country, addressing a variety of groups. A question I would often hear at the end of such addresses is this: how come we never hear things like this from the pulpit? That question could have come from a college professor or from a professional environmentalist. I suspect that that professor and that environmentalist *might* have heard such matters discussed in their home parishes, but that they might not have been prepared to take them to heart, then and there. As a pastoral practitioner for many years, I am well aware of the difficulties preaching what one believes in this respect, not to speak of practicing what one preaches, especially in these times of global emergency.

Faced with this kind of world, preachers need all the help they can get. This is why the volume edited by David Rhoads, *Earth & Word: Classic Sermons on Saving the Planet*,[22] to which I contributed one of the 36 sermons collected there, is so important for the whole ecumenical Church. It's not easy to preach from the Church's lectionary and, at the same time, to address both the themes of ecology and justice effectively. The sermons

---

21. Santmire, *Nature Reborn*.
22. Rhoads, ed., *Earth and Word*.

might well be biblical, incisive, and well-delivered, but the congregations' readiness to listen and then respond may not be sufficient, given everything else that is on their minds. The "anguish of preaching" that Joseph Sittler memorably talked about in 1966 is still with us, perhaps more so than six decades ago.[23]

In the last few years, therefore, I have come to this conclusion, with ever-firmer conviction: that the theology of the kind that comes to expression in *Nature Reborn,* in my speeches on the stump, and in my preaching, while good and true and beautiful (of course), is not enough. Lutherans like myself have invested enormously in the theology of the Word, following Luther himself (Luther once referred to a church building, as we have seen, as a "mouth house"). But something, it has appeared to me, has been missing.

Coming Home to the Liturgy

Which brings me to consider what I now think is the culminating stage of my own theological trajectory, to which this book belongs, signaled by two earlier volumes that I have written in my retirement: *Ritualizing Nature: Renewing Christian Liturgy in a Time of Crisis* and *Before Nature: A Christian Spirituality.*[24] I did have chapters on ritual and spirituality in *Nature Reborn,* but these more recent books represent a much more thorough immersion in those deep waters. I will consider each book, in turn.

Throughout the more than fifty years of my vocational trajectory, I have, as will be apparent by now, always thought things through as a practitioner, not just as a scholar, nor just as some kind of peripatetic theological stump-speaker. If, indeed, I were to call forth one image of my vocational trajectory, before all others, I would see myself preaching and officiating at the Eucharist. The liturgy of the Church, and my calling to preside over that liturgy, has always been at the heart of my theological life.

It began with a very good grounding, as a protege of one of the preeminent American pastoral and liturgical practitioners of the second half of the last century, the Rev. Dr. Henry Horn in Cambridge, Massachusetts,

---

23. Sittler, *Anguish of Preaching.* Part of the reason why the anguish of preaching is so deeply seated in our times is that we are facing an apocalyptic situation, on the one hand, and the reemergence of a highly publicized kind of apocalyptic preaching in some Protestant Churches, on the other. It's exceedingly difficult, therefore, to interpret apocalyptic texts without sounding like some standard end-of-the-world popular preacher. For some much-needed help in this respect, see Rossing, "The World is About to Turn," 140–59.

24. Santmire, *Ritualizing Nature*; and Santmire, *Before Nature.*

the pastor with whom I first talked, during my college years, about my theological and spiritual troubles dealing with the Holocaust and who later mentored me, when I served as his pastoral assistant at Harvard. He taught me, mostly by example, how it was possible for celebrants to live their way into the liturgy, not just cognitively, but much more so, spiritually and existentially. Later, at Wellesley College, I struggled to make available the deep claims of the liturgy in the College Chapel, in the midst of a sixties and seventies culture that made it easy for faculty and students to assume that inherited forms of ritual were either irrelevant or counterrevolutionary.

During my ensuing thirteen years as an inner-city pastor in the then fourth-poorest city in the country, Hartford, Connecticut, I did preside over the transformation of a congregation from a white German-ethnic community to a racially mixed neighborhood church, and I did become the godfather of an Alinsky-style neighborhood organization. But I invested still more energy encouraging that congregation allow itself to be claimed by the historic liturgy of the Church.

Likewise for my ministry for seven years in a large, historic downtown congregation in Akron, Ohio, Holy Trinity Lutheran Church. Housed in a beautiful and spacious neogothic building, blessed with one of the great pipe organs in the country, and proud of its venerable Lutheran heritage, that congregation, or at least its leaders at that time, *knew* what their worship *should* be. Call this the gothic dream of middle-American Protestantism—a beautiful dream, but, in my view, much too vulnerable to the forces of acculturalization, much too predisposed to foster a congregation that was what Churchill said of the Anglican Church of his day, "the Tory party at prayer." It was a struggle, therefore, to introduce that congregation to some of the major reforms that emerged from the movement for liturgical renewal a half century before, not to speak of themes from liberation theology. But I relished that struggle.

Some things, I suppose, never change. I began my life in the Church as a sacramental Lutheran and I am now concluding my life in the Church as a sacramental Lutheran, although hopefully of a higher order. My ministry, both as a practitioner and as a writing theologian, has always been shaped—insofar as it has been given me so to do—by Luther's understanding of the real presence of Jesus Christ in Word and Sacrament and in the people of Christ ministering to each other and to the world, for the sake of what Luther liked to call "the forgiveness of sins, life, and salvation." This is why I think that my book *Ritualizing Nature* represents one culmination of my whole vocational trajectory.

Celebrating Nature by Faith

It is very much an experimental work. It begins correlatively, in Tillichian fashion, probing questions that readers are actually thinking about, such as: why would anyone in today's postmodern world even want to think about exploring the claims of the liturgy? The book then seeks to bind together the insights of liturgical studies, on the one hand, and the theology of ecology and justice, on the other. Hence I was grateful and also encouraged when two prominent Lutheran theologians, whose works I considered above, the liturgical scholar, Gordon Lathrop, and the social ethicist, Larry Rasmussen, both endorsed the project.

What does "ritualizing nature" mean? Consider this premise, that liturgy is the Church's mode of identity-formation. Not theology, as such. Not merely preaching, which is theology as personal address (the *viva vox evangelii*). No, the Church's ritual, its liturgy, makes all the difference. Such a premise reflects a wide range of cultural studies, which show that ritual, more generally, is the human mode of identity-formation. Erik Erikson, for example, argued that without positive rituals, the human infant would not develop what Erikson called basic trust and therefore would not be able to grow into psychological maturity.[25]

An example: morning after morning, a parent comes into a child's room and smiles at the child. That ritual inculcates what Erikson called basic trust. Analogously, for Christian ritual: when Christians "do this," as the Lord commanded, when they break bread and drink wine together, they are practicing, embodying, becoming habituated to, the self-giving love of Christ, who "on the night in which he was betrayed" said, "do this in remembrance of me" (see 1 Cor 11:23–24). Christian ritual thereby forms Christian character, which, in turn, shapes Christian action in the world.

But what if Christian ritual happens to be highly spiritualized, encouraging participants to rise above the world of material things in order to commune with God in heaven? Will not the result be habits of disinterest in nature, perhaps even habits of domination of nature? What if Christian preaching in the context of such a ritual is driven by a theology of self-esteem or even by the quest for the forgiveness of sins alone? Will not the result of such liturgical preaching tend to be a powerful focus on the individual believer and on his or her needs?

What about a Sunday ritual that is predicated on leading Christians into a personal relationship with Jesus Christ (this is not only an American-Evangelical theme; notably, it also flows from the heart of the Lutheran

---

25. See Erikson, "The Development of Ritualization."

tradition, as the music of Bach often shows)? Will those believers then be inclined only to contemplate Jesus Christ as their personal savior? Must not liturgical preaching, indeed, be driven by a vision of the whole economy of God, from the mysteries of creation to the ineffable consummation of all things, with Jesus Christ proclaimed as the redeemer and savior of all things, by making peace by the blood of his Cross, according to the imaginative immenseness of the faith that comes to expression in Col 1:15ff.?

But preaching cannot be our only concern, in this respect, or even our main concern. In *Ritualizing Nature*, I argue that *the whole liturgy must be right*, if the Church's ecological and justice praxis is to be right. Practice will not make perfect. But practice—if it is good practice—will in all likelihood make possible.

Consider the shape of the eucharistic prayer as a case in point. The tradition on the side of the Protestant Reformation has been to minimize the use of a full Trinitarian eucharistic prayer, for fear of introducing themes of sacrifice that contradict the Gospel of the free grace of God. The tradition on the side of the Catholic Reformation has also been to minimize the use of a full Trinitarian eucharistic prayer, in favor of the "words of institution," spoken silently, as the essence, the consecrating moment, of the Mass. The theology of penance, inherited from the late Middle Ages—and with that theology, a solitary kind of emphasis on the forgiveness of sins—has thus shaped the traditions of both the Protestant Reformation and the Catholic Reformation.

This has meant that a transaction addressing the individual believer's interiority has moved to the center of the liturgical experience: it is all "for me" (*pro me*). Consider, in contrast, how the post-Vatican II reemphasis on the full eucharistic prayer, in both Protestant and Catholic circles, has broadened eucharistic horizons of meaning. Now the ecumenical Church gives thanks for the fullness of God's creative and redemptive activity, from the Alpha to the Omega of creation-history, with the individual believer receiving Grace within that universal setting. Call this the Cosmic Mass (Teilhard de Chardin) of the ecumenical Church. In that context, we are shaped by what is done and that "what" is nothing less than what is being done by God in, with, and under the whole creation and in, with, and under the communal ritual of the Church, in particular.

The Protestant-Reformation and the Catholic-Reformation Liturgies thus have tended to shape Christians mainly for their solitary spiritual struggles, while the Post-Vatican II Liturgy tends to shape them also for their communal involvement in God's wondrous and sometimes alienating works within the whole creation, as well as in God's marvelous and miraculous works within the Church as a ritual community.

In this way, *Ritualizing Nature* brings together for me, one more time, and for me in a most gratifying and self-conscious fashion, those two themes that have preoccupied me the most throughout the course of my whole theological trajectory, ecology and justice, each one and both together driven by the promise of the Gospel, announced and formed in the Church's ritual practices. As a liturgical practitioner for more than five decades, I now realize that I have been seeking to discover and embody this kind of unified theological vision my whole vocational life, in my preaching, my teaching, my writing, my counseling, my officiating, my domestic life, and my public witness.

Professing the Promise of an Ecological Spirituality

This brings me to the work which is a companion volume for my book on the Liturgy: *Before Nature: A Christian Spirituality*. I have here eagerly entered into an arena where angels fear to tread. Everybody these days, it seems, is interested in spirituality. Why not me?

The book is written for two audiences, first for the increasingly large body of spiritual seekers, both outside and within our churches, who say they are more interested in spirituality than religion, second for those numerous theologians and pastoral practitioners who are eager to be in conversation with those seekers.

The title has two meanings. I stand *before* nature, as I have my whole life, engaged with its beauties, its immensities, its terrors, and its increasingly caustic desecrations by human folly, and I contemplate nature charged with the glories of God. My faith in God, however, has always been grounded in the life of the Church existentially, *before* that kind of encounter with nature. Indeed, as I understand it, my faith begins with my baptism and with what I have called my baptismal mysticism. Hence the centrality of the Trinity in my spirituality of nature: "in the name of the Father and of the Son and of the Holy Spirit."

Before Nature is a kind of meditation on the Trinity in, with, and under the whole cosmos. This approach, of course, is culturally problematic in these times, given the well-recognized liabilities of patriarchal thought and practice. But in *Before Nature* I make every effort to avoid patriarchal readings of the mystery of the Trinity. For me, God is Giver, Gift, and Giving; and the Father, in particular, I envision, with Moltmann, as the "motherly Father" who suffers in the Son and with the Son, as the Crucified God.

Presupposing such understandings, in *Before Nature* I invite readers to practice what I call the Trinity Prayer: "Lord Jesus Christ, have mercy on me. Praise Father, Son, and Holy Spirit. Come Holy Spirit, come and reign." And I explore the cosmic ministries of what Irenaeus called the "two hands of the Father," the Son and the Spirit.

I do all this by taking the readers with me to several particular places, where my own spirituality has been or is grounded, among them: scything and gardening in a rural Maine setting, walking along the Charles River in Cambridge, Massachusetts and sauntering through the rich milieu of a great arboretum in the same city, entering into liturgical perambulations in a Cambridge monastery and an inner-city Boston parish, experiencing work (long ago) in a steel plant outside Buffalo, New York (when I almost lost my life), and contemplating the wonders of the great Niagara River and its Falls (also long ago). All the time, I seek to "pray without ceasing," invoking the Trinity Prayer subliminally, if not vocally, whenever and wherever possible.

Before Nature incorporates testimony to all the dimensions of God's good creation that have so captivated me for so long, the Church's liturgical practices and my own practices of prayer, the city and the countryside, the stands of surviving wilderness areas on this good Earth and the virtually infinite reaches of the cosmos beyond, all shaped, I believe, by the ministry of the Cosmic Christ and all driven by the energies of Cosmic Spirit.

Living in an Ecumenical World

And more. My quest for a unified theological vision was predicated on a longstanding commitment of my own, which I hope will come as no surprise to any attentive reader at this point, but which I want to celebrate explicitly before I conclude this narrative. I began my theological studies when ecumenism was on the rise and I have never left that world behind. All along, my own theological roots have been richly nourished by a wide variety of ecumenical witnesses.

I have listened carefully to voices from the traditions of the Methodists, thinkers such as John Cobb and James Nash. The Reformed tradition has been a major influence in my thinking, from Calvin himself to Karl

Barth and then Juergen Moltmann. Moreover, the witness of what some used to think of as the twentieth century particularists or, better, the liberationists, has been very much part of my theological world, as well, especially that great champion of Black Theology, James Cone, and representatives of what has now become a full-fledged theological movement, ecofeminism. The most influential, for me: Mary Daly, Rosemarie Radford Ruether, Sallie McFague, and, most recently, Elizabeth Johnson.

The theology of Vatican II has found a place in my heart, too, particularly its liturgical expressions, as I have already indicated. I have written appreciatively about a variety of other Catholic witnesses in *The Travail of Nature*, as well, among them St. Augustine, the Celtic saints, St. Francis, and Pierre Teilhard de Chardin. Most recently, like many Christians around the world, I have been stunned, as I said at the outset of this book, by the insight and the power of Pope Francis' essentially Franciscan encyclical in 2015, *Laudato Si': On Caring for Our Common Home*. In a certain sense, that theological document has taken all the ecological air out of the ecumenical room, and I celebrate that.

In *Laudato Si'*, Pope Francis has almost seamlessly joined the two causes that have long preoccupied me, ecology and justice. So the liberation theologian Leonardo Boff has rightly called *Laudato Si'* "the Magna Carta of integral ecology," because it is responsive both to the cry of the earth and to the cry of the poor.[26]

But however much I have been influenced by the riches of many Christian traditions, I was born a Lutheran and, for better or for worse, once born I have remained. As I hope will be apparent by now, this identity has sometimes been a struggle for me, especially in the wake of the Lutheran tradition's scandalous response to the phenomenon of National Socialism in Germany. I have also found the Lutheran tradition wanting insofar as it has not until very recently offered theological resources to help the Church respond to our global ecojustice crisis and to the challenges posed by the natural sciences in our time, especially evolutionary biology and cosmological physics.

With respect to Luther, I have made an intense effort to show how he himself was, as it were, an ecological theologian before his time (see Chapter 2 above) and how a number of late twentieth century Lutheran theologians in the U.S. made substantive—if not always widely recognized—contributions to ecological theology (see Chapter 3 and Chapter 4

---

26. Boff, "The Magna Carta of Integral Ecology."

above). In so doing, I have also tried to identify the two ways of contemporary Lutheran thinking about nature, one of which is ecological in scope, the other of which is not. As a matter of course, I have understood myself as a champion of that first way of thinking, which I have called Lutheran theological maximalism—over against a minimalist, exclusive Lutheran focus on justification by faith alone.

Such was my theological trajectory in the past few years. Beyond my ecumenical concerns, I have always tried to take my own theological tradition seriously as well as critically and have always tried to take responsibility for it, as best as I could. There is doubtless much more to be said about Reformation theology from the perspective of an ecumenical ecological theology, especially about Lutheran social ethics, positively and negatively, but I think that at this point in my life I have said everything that I can say.

Coda: Behold the Lilies and Celebrating Nature by Faith

My most recent books, *Behold the Lilies: Jesus and the Contemplation of Nature—a Primer*[27] and the volume in your hands that I am now beginning to bring to a close, *Celebrating Nature by Faith: Studies in Reformation Theology in an Era of Global Emergency,* illustrate how my longstanding commitment to kerygmatic theology is coming to a certain termination (*finis*) as I write this (I am now in my mid-eighties), but also to its fulfilment (*telos*), at least in my own mind. For me, kerygmatic theology entails, among other things, both personal testimony and historical roots. *Behold the Lilies* accents the former. *Celebrating Nature by Faith* accents the latter, although not without some personal testimony along the way, culminating in this chapter.

Behold the Lilies is something of a paradox. It says everything I have to say personally about the Christian story—the kerygma—without saying everything. Depending on one's angle of vision, this primer skims the surface or plumbs the depths of the theocosmocentrism that has shaped my own theology for as long as I can remember. But it does so not discursively, but cryptically and enigmatically, as a response to what I take to be a *command* of Jesus, as I adopt the translation of that New Testament text by Joseph Sittler: "Behold the lilies."

---

27. Santmire, *Behold the Lilies.*

Celebrating Nature by Faith

In my view, this command of Jesus is non-negotiable for those who are seeking to be his disciples, like other commands by Jesus, such as "Follow me" (Matt 4:19). *Behold the Lilies* also reads those words as giving expression to what I have long called—once again, in these pages—an I–Ens relationship with nature, which I have described in conversation with Martin Buber's depictions of I–Thou and I–It relationships. Throughout, too, *Behold the Lilies* addresses my passionate engagement with both ecological and justice issues over the years.

Celebrating Nature by Faith, in contrast, is, as will have been evident from the very beginning, a discursive, historical book for the most part. It explores the kerygma by digging deeply into the Scriptures and into Luther's theology, then into a single chapter of Reformation theology in the U.S., extending for some fifty years, beginning in 1961 with Joseph Sittler's scintillating address to the World Council of Churches and concluding, perhaps improbably, with these down-to-earth autobiographical explorations.

With this book's final chapter, then, I bring to conclusion a particular theological narrative, an account of a single ecotheological trajectory, as seen at the ground level. I hope that the explorations and remembrances in this chapter, filtered through the dark netting of my own subjectivity as they all have been, will nevertheless be helpful for anyone who wants to understand the ebb and flow of an ecumenically oriented Reformation ecotheology in the U.S. since the middle of the last century, as numbers of us, in various ways, committed ourselves in those decades to celebrate nature by faith, in pursuit of an ecological reformation of Christianity.

Will the reader, therefore, at the end of this book, having taken one step backward with me to explore a theological legacy of more than fifty years in the making, now be inspired to take two theological steps forward to celebrate nature by faith on his or her own, more resolutely and with a fresh sense of promise? Godspeed.

Bibliography

Adams, Edward. *Constructing the World: A Study in Paul's Cosmological Language*. Studies of the New Testament and Its World. Edinburgh: T. & T. Clark, 2000.
Albrecht, Paul, ed. *The World Council of Churches Conference on Faith, Science, and Technology*. 2 vols. Geneva: World Council of Churches, 1979.
Althaus, Paul. *The Theology of Martin Luther*. Translated by Robert C. Schultz. Minneapolis: Fortress, 1966.
Anderson, Bernhard W. "Creation and Ecology." In *Creation in the Old Testament*, edited by Bernhard W. Anderson, 152–78. Issues in Religion and Theology 6. Philadelphia: Fortress, 1984.
———, ed. *Creation in the Old Testament*. Issues in Religion and Theology 6. Philadelphia: Fortress, 1984.
———. "Introduction: Mythopoeic and Theological Dimensions of Biblical Creation Faith." In *Creation in the Old Testament*, edited by Bernhard W. Anderson, 1–24. Issues in Religion and Theology 6. Philadelphia: Fortress, 1984.
Bagchi, David, and David C. Steinmetz, eds. *The Cambridge Companion to Reformation Theology*. Cambridge Companions to Religion. Cambridge: Cambridge University Press, 2004.
Bakken, Peter W. "The Ecology of Grace: Ultimacy and Environmental Ethics in Aldo Leopold and Joseph Sittler." PhD diss., University of Chicago, 1991.
Bakken, Peter W., et al. *Ecology, Justice, and Christian Faith: A Critical Guide to the Literature*. Bibliographies and Indexes in Religious Studies 36. Westport, CT: Greenwood, 1995.
Barth, Karl. *The Word of God and the Word of Man*. Translated by Douglas Horton, 1928. Reprint, Gloucester, MA: Smith, 1957.
Bauckham, Richard J. *The Bible and Ecology: Rediscovering the Community of Creation*. Sarum Theological Lectures. London: Darton, Longman & Todd, 2010.
Blumenberg, Hans. "Light as a Metaphor for Truth: At the Preliminary Stage of Philosophical Concept Formation." In *Modernity and the Hegemony of Vision*, edited by David Michael Levin, 30–62. Berkeley: University of California Press, 1993.
Bornkamm, Heinrich. *Luther's World of Thought*. Translated by Martin K. Bertram. St. Louis: Concordia, 1965.
Bouma-Prediger, Steven. *For the Beauty of the Earth: A Christian Vision for Creation Care*. Engaging Culture. Grand Rapids: Baker Academic, 2001.
———. *The Greening of Theology: The Ecological Models of Rosemary Radford Ruether, Joseph Sittler, and Jürgen Moltmann*. American Academy of Religion Academy Series 91. Atlanta: Scholars, 1995.

Bibliography

———. "Sittler the Pioneering Ecological Theologian." In *Evocations of Grace: The Writings of Joseph Sittler on Ecology, Theology, and Ethics*, edited by Steven Bouma-Prediger and Peter Bakken, 223–33. Grand Rapids: Eerdmans, 2000.

Braaten, Carl E., and Robert W. Jenson, eds. *Christian Dogmatics*. 2 vols. Philadelphia: Fortress, 1984.

Brett, Mark G. "Earthing the Human in Genesis 1–3." In *The Earth Story in Genesis*, edited by Norman C. Habel and Shirley Wurst, 73–86. Earth Bible 2. Cleveland: Pilgrim, 2000.

Brown, William P. *The Ethos of the Cosmos: The Genesis of Moral Imagination in the Bible*. Grand Rapids: Eerdmans, 1999.

Brueggemann, Walter. *Genesis*. Interpretation. Atlanta: John Knox, 1982.

Brunner, Emil. *Revelation and Reason*. Translated by Olive Wyon. Philadelphia: Westminster, 1946.

Buber, Martin. *I and Thou*. 2nd ed. Scribner Library. New York: Scribner, 1958.

Buller, Cornelius A. *The Unity of Nature and History in Pannenberg's Theology*. Lanham, MD: Rowman & Littlefield, 1996.

Carson, Rachel. *Silent Spring*. Boston: Houghton Mifflin, 1962.

Chidester, David. *Word and Light: Seeing, Hearing, and Religious Discourse*. Urbana: University of Illinois Press, 1992.

Childs, Brevard S. *Biblical Theology of the Old and New Testaments: Theological Reflections on the Christian Bible*. Minneapolis: Fortress, 1993.

Childs, James M., Jr. "Nothing Less than Everything: Thoughts on a Sittler Legacy." *Dialog* 46 (2007) 104–11.

Cobb, John B., Jr. *Is It Too Late? A Theology of Ecology*. Beverly Hills, CA: Bruce, 1972.

Conradie, Ernst M., ed. *Creation and Salvation: Dialogue on Abraham Kuyper's Legacy for Contemporary Ecotheology*. 2 vols. Studies in Reformed Theology. Zurich: Lit, 2012.

Dahill, Lisa E. "Rewilding Christian Spirituality: Outdoor Sacraments and the Life of the World." In *Eco-Reformation: Grace and Hope for a Planet in Peril*, edited by Lisa E. Dahill and James B. Martin-Schramm, 177–96. Eugene, OR: Cascade Books, 2016.

Dahill, Lisa E., and James B. Martin-Schramm, eds. *Eco-Reformation: Grace and Hope for a Planet in Peril*. Eugene, OR: Cascade Books, 2016.

Daly, Mary. *Beyond God the Father: Toward a Philosophy of Women's Liberation*. Boston: Beacon, 1973.

de Gruchy, John W. "A Concrete Ethic of the Cross: Interpreting Bonhoeffer's Ethics in North America's Backyard." In "Fidelity to Earth: A Festschrift in Honor of Larry Rasmussen," edited by Daniel T. Spencer and James B. Martin-Schramm, special issue, *Union Seminary Quarterly Review* 58/1–2 (2004) 33–45.

Dowey, Edward A. *The Knowledge of God in Calvin's Theology*. New York: Columbia University Press, 1952.

Drummy, Michael F. *Being and Earth: Paul Tillich's Theology of Nature*. Lanham, MD: University Press of America, 2000

Dubos, René. *Reason Awake: Science for Man*. New York: Columbia University Pres, 1970.

Eckhart, Meister. *Meister Eckhart, German Sermons & Treatises*. Translated with introduction and notes by M. O'C. Walshe. 2 vols. London: Watkins, 1979–1981.

Edwards, Denis. *Christian Understandings of Creation: The Historical Trajectory*. Christian Understandings. Minneapolis: Fortress, 2017.

———. *Jesus and the Cosmos*. New York: Paulist, 1991.

Bibliography

———. *Jesus the Wisdom of God: An Ecological Theology*. Ecology and Justice. Maryknoll, NY: Orbis, 1995.

Edwards, Mark U., Jr. "Luther's Polemical Controversies." In *The Cambridge Companion to Reformation Theology*, edited by David Bagchi and David C. Steinmetz, 192–208. Cambridge Companions to Religion. Cambridge: Cambridge University Press, 2003.

Evangelical Lutheran Church in America. *Evangelical Lutheran Worship*. Minneapolis: Augsburg Fortress, 2006.

Fletcher, Joseph. *Situational Ethics: The New Morality*. London: SCM, 1970.

Fowler, Robert Booth. *The Greening of Protestant Thought*. Chapel Hill: University of North Carolina Press, 1995.

Francis I, Pope. *Laudato Si': On Care of Our Common Home*. Vatican City: Vaticana, 2015.

Frietheim, Terence E. "Creator, Creature, and Co-Creation in Genesis 1–2." In *All Things New: Essays in Honor of Roy A. Harrisville*, edited by Arland J. Hultgren et al., 11–20. Word & World Supplement Series 1. St. Paul: Word & World, Luther Northwestern Theological Seminary, 1992.

———. *God and World in the Old Testament: A Relational Theology of Creation*. Nashville: Abingdon, 2005.

———. "Is Genesis 3 a Fall Story?" *Word & World* 14 (1994) 144–53.

———. "Nature's Praise of God in the Psalms." *Ex Auditu* 3 (1987) 16–30.

———. "The Plagues as Ecological Signs of Historical Disaster." *Journal of Biblical Literature* 110 (1991) 385–96.

———. "The Reclamation of Creation: Redemption and Law in Exodus." *Interpretation* 45 (1991) 354–65.

———. *The Suffering of God: An Old Testament Perspective*. Overtures to Biblical Theology 14. Philadelphia: Fortress, 1984.

Gottlieb, Roger S., ed. *The Oxford Handbook of Religion and Ecology*. Oxford: Oxford: University Press, 2006.

Graham, William A., ed. *Beyond the Written Word: Oral Aspects of Scripture in the History of Religion*. Cambridge: Cambridge University Press, 1987.

———. "Hearing and Seeing: The Rhetoric of Martin Luther." In *Beyond the Written Word: Oral Aspects of Scripture in the History of Religion*, 141–54. Cambridge: Cambridge University Press, 1987.

Green, Joel B. "Scripture and Theology: Failed Experiments, Fresh Perspectives." *Interpretation* 56 (2002) 5–20

Gregersen, Niels Henrik. "Deep Incarnation and Kenosis: In, With, and Under, and As: A Response to Ted Peters." *Dialog* 52 (2013) 251–62.

———. "Grace in Nature and History: Luther's Doctrine of Creation Revisited." *Dialog* 44 (2005) 19–29.

———, ed. *Incarnation: On the Scope and Depth of Christology*. Minneapolis: Fortress, 2015.

Grossberg, Daniel. "Nature, Humanity, and Love in Song of Songs." *Interpretation* 59 (2005) 229–43.

Habel, Norman C. *The Land Is Mine: Six Biblical Land Ideologies*. Overtures to Biblical Theology. Minneapolis: Fortress, 1995.

———, ed. *Readings from the Perspective of Earth*. Earth Bible 1. Cleveland: Pilgrim, 2000.

Habel, Norman C., et al., eds. *The Season of Creation: A Preaching Commentary*. Minneapolis: Fortress, 2011.

BIBLIOGRAPHY

Hall, Douglas John. *Imagining God: Dominion as Stewardship*. 1986. Reprint, Eugene, OR: Wipf & Stock, 2004.

Hazelton, Roger. "The Nature of Christian Paradox." *Theology Today* 6 (1949) 324–35.

Hefner, Philip. "Can a Theology of Nature Be Coherent with Scientific Cosmology?" *Dialog* 30 (1991) 267–72.

———. "The Creation." In *Christian Dogmatics*, edited by Carl E. Braaten and Robert W. Jenson, 1:265–358. 2 vols. Philadelphia: Fortress, 1984.

———. *The Human Factor: Evolution, Culture, and Religion*. Theology and the Sciences. Minneapolis: Fortress, 1993.

———. "Nature's History as Our History: A Proposal for Spirituality." in *After Nature's Revolt: Eco-Justice and Theology*, edited by Dieter T. Hessel, 171–83. Minneapolis: Fortress, 1992.

———. *Technology and Human Becoming*. Facets. Minneapolis: Fortress, 2003.

Hendel, Kurt K. "*Finitum Capax Infiniti*: Luther's Radical Incarnational Perspective." *Seminary Ridge Review* 10/2 (2008) 20–35.

Hendrix, Scott H. "Luther." In *The Cambridge Companion to Reformation Theology*, edited by David Bagchi and David C. Steinmetz, 39–56. Cambridge Companions to Religion. Cambridge: Cambridge University Press, 2005.

Hessel, Dieter T., ed. *After Nature's Revolt: Eco-Justice and Theology*. Minneapolis: Fortress, 1992.

Hessel, Dieter T., and Larry L. Rasmussen. *Earth Habitat: Eco-Injustice and the Church's Response*. Minneapolis: Fortress, 2001.

Hiebert, Theodore. "Re-imaging Nature: Shifts in Biblical Interpretation." *Interpretation* 50 (1996) 36–46.

———. "Rethinking Traditional Approaches to Nature in the Bible." In *Theology for Earth Community: A Field Guide*, edited by Dieter T. Hessel, 23–30. Ecology and Justice. Maryknoll, NY: Orbis, 1996.

———. *The Yahwist's Landscape: Nature and Religion in Early Israel*. New York: Oxford University Press, 1996.

Hoffmann, Bengt R. *Luther and the Mystics: A Re-examination of Luther's Spiritual Experience and His Relationship to the Mystics*. Minneapolis: Augsburg, 1976.

———, trans. *The Theologia Germanica of Martin Luther*. Introduction and commentary by Bengt R. Hoffmann. Preface by Bengt Hagglund. The Classics of Western Spirituality. New York: Paulist, 1980.

Horrell, David G., et al. *Greening Paul: Rereading the Apostle in a Time of Ecological Crisis*. Waco, TX: Baylor University Press, 2010.

Hopkins, Gerard Manley. "God's Grandeur." *Poetry Foundation* (website). https://www.poetryfoundation.org/poems/44395/gods-grandeur/.

Howells, Edward. "Apophatic Spirituality." In *The New Westminster Dictionary of Christian Spirituality*, edited by Philip Sheldrake, 117–19. Louisville: Westminster John Knox, 2005.

Hunsinger, George. "Aquinas, Luther, and Calvin: Toward a Chalcedonian Resolution." In *The Gift of Grace: The Future of Lutheran Theology*, edited by Niels Henrik Gregersen et al., 181–93. Minneapolis: Fortress, 2005.

Janzen, J. Gerald. "Creation and the Human Predicament in Job." *Ex Auditu* 3 (1987) 45–53.

Jenkins, Willis. *Ecologies of Grace: Environmental Ethics and Christian Theology*. New York: Oxford University Press, 2008.

BIBLIOGRAPHY

Jensen, Franklin L., and Cedric W. Tilberg, eds. *The Human Crisis in Ecology.* Christian Social Responsibility. New York: Board of Social Ministry, Lutheran Church in America, 1972.

Jenson, Robert W. *America's Theologian: A Recommendation of Jonathan Edwards.* New York: Oxford University Press, 1988.

———. "How the World Lost Its Story." *First Things* 38 (October 1993) 19–24.

Jorgenson, Kiara A. *Ecology of Vocation: Recasting Calling in a New Planetary Era.* New York: Fortress Academic, 2020.

Kahl, Brigitte. "Fratricide and Ecocide: Rereading Genesis 2–4." In *Earth Habitat: Eco-Injustice and the Church's Response,* edited by Dieter T. Hessel and Larry L. Rasmussen, 53–70. Minneapolis: Fortress, 2001.

Kaufman, Gordon D. "A Problem for Theology: The Concept of Nature." *Harvard Theological Review* 65 (1972) 337–66.

Kleckley, Russell. *Omnes Creaturae Sacramenta: Creation, Nature, and World View in Luther's Theology of the Lord's Supper.* Dissertation zu Erlangung der Doktorwürde der Ludwig-Maximilians-Universität München, Evangelisch-Theologische Fakultät. Columbia, SC, privately published, 1990.

Knierim, Rolf P. "Cosmos and History in Israel's Theology." In *The Task of Old Testament Theology: Substance, Method, and Cases,* 177–224. Grand Rapids: Eerdmans, 1995.

Koerner, Joseph Leo. *The Reformation of the Image.* Chicago: University of Chicago Press, 2004.

Landes, George M. "Creation and Liberation." In *Creation in the Old Testament,* edited by Bernhard W. Anderson, 135–51. Issues in Religion and Theology 6. Philadelphia: Fortress, 1984.

Lane, Belden C. *Ravished by Beauty: The Surprising Legacy of Reformed Spirituality.* New York: Oxford University Press, 2011.

Lathrop, Gordon W. *Holy Ground: A Liturgical Cosmology.* Minneapolis: Fortress, 2003.

Levenson, Jon D. *Creation and the Persistence of Evil: The Jewish Drama of Divine Omnipotence.* San Francisco: Harper & Row, 1988.

———. *Sinai and Zion: An Entry into the Jewish Bible.* San Francisco: Harper & Row, 1985.

Levin, David Michael, ed. *Modernity and the Hegemony of Vision.* Berkeley: University of California Press, 1993.

Lindberg, Carter. "Luther's Struggle with Social-Ethical Issues." In *The Cambridge Companion to Reformation Theology,* edited by David Bagchi and David C. Steinmetz, 165–78. Cambridge Companions to Religion. Cambridge: Cambridge University Press, 2005.

Loefgren, David. *Die Theologie der Schoepfung bei Luther.* Forschungen zur Kirchen- und Dogmengeschichte 10. Göttingen: Vandenhoeck & Ruprecht, 1960.

Lohse, Bernhard. *Martin Luther's Theology: Its Historical and Systematic Development.* Translated and edited by Roy A. Harrisville. Minneapolis: Fortress, 1999.

Luther, Martin. *Luther's Works,* ed. Jaroslav Pelikan and Helmut Lehman. St. Louis: Concordia, multiple dates.

———. "Small Catechism." In *The Book of Concord,* edited by Theodore G. Tappert, 337–56. Philadelphia: Fortress, 1959.

———. *Theologica Germanica.* Translated by Bengt R. Hoffman. Mahwah, NJ: Paulist, 1980.

———. *Werke: Kritische Gesammtausgabe.* Weimarer Ausgabe, multiple dates.

Bibliography

Lutheran Church in America. "The Human Crisis in Ecology." Social Statements, Lutheran Church in America. New York: Division for Mission in North America, Lutheran Church in America, 1972.

Lutz, Paul E., and H. Paul Santmire. *Ecological Renewal.* Edited by William H. Lazareth. Confrontation Books. Philadelphia: Fortress, 1972.

MacKinnon, Mary Heather, and Moni McIntyre, eds. *Readings in Ecology and Feminist Theology.* Kansas City: Sheed & Ward, 1995.

McDaniel, Jay. "'Where is the Holy Spirit Anyway?' Response to a Skeptical Environmentalist." *Ecumenical Review* 42/2 (1990) 162–74.

McFague, Sallie. *The Body of God: An Ecological Theology.* Minneapolis: Fortress, 1993.

———. *Metaphorical Theology: Models of God in Religious Language.* Philadelphia: Fortress, 1982.

McGrath, Alister. *Luther's Theology of the Cross: Martin Luther's Theological Breakthrough.* New York: Blackwell, 1985.

McKibben, Bill. *The End of Nature.* New York: Random House, 1989.

Miles, Margaret R. *Image as Insight: Visual Understanding in Western Christianity and Secular Culture.* Boston: Beacon, 1985.

———. *Image as Insight: Visual Understanding in Western Christianity and Secular Culture.* 1985. Reprint, Eugene, OR: Wipf & Stock, 2006.

Moe-Lobeda, Cynthia D. "Christian Ethics toward Earth-Honoring Faiths." *Union Theological Seminary Quarterly Review* 58/1–2 (2004) 132–50.

———. *Healing a Broken World: Globalization and God.* Minneapolis: Fortress, 2002.

———. *Resisting Structural Evil: Love as Ecological-Economic Vocation.* Minneapolis: Fortress, 2013.

Moltmann, Jürgen. *The Trinity and the Kingdom: The Doctrine of God.* Translated by Margaret Kohl. San Francisco: Harper &Row, 1981.

———. *The Way of Jesus Christ: Christology in Messianic Dimensions.* Translated by Margaret Kohl. Minneapolis: Fortress, 1993.

Morgan, David. *Visual Piety: A History and Theory of Popular Religious Images.* Berkeley: University of California Press, 1998.

Murphy, George L. *The Cosmos in Light of the Cross.* Harrisburg, PA: Trinity, 2003.

Murphy, George L., et al. *Cosmic Witness: Commentaries on Science/Technology Themes.* Lima, OH: CSS, 1996.

Murray, Shirley Erena. "Touch the Earth Lightly." In *Evangelical Lutheran Worship*, hymn 739. Minneapolis: Augsburg Fortress, 2006.

Nash, James A. *Loving Nature: Ecological Integrity and Christian Responsibility.* Nashville: Abingdon, 1991.

———. "Toward an Ecological Reformation of Christianity." *Interpretation* 50 (1996) 5–15.

Oberman, Heiko A. *Luther: Man between God and the Devil.* Translated by Eileen Walliser-Schwarzbart. New Haven: Yale University Press, 1989.

Orr, David W. *Down to the Wire: Confronting Climate Collapse.* New York: Oxford University Press, 2012.

Pelikan, Jaroslav. *Luther the Expositor: Introduction to the Reformer's Exegetical Writings.* St. Louis: Concordia, 1959.

Pesch, Otto H. "Existential and Sapiential Theology—The Theological Confrontation between Luther and Thomas Aquinas." In *Catholic Scholars Dialogue with Luther*, compiled by Jared Wicks, 61–81. Chicago: Loyola University Press, 1970.

Bibliography

Peters, Ted. *Anticipating Omega: Science, Faith, and Our Ultimate Future*. Religion, Theology, and Natural Science 7. Göttingen: Vandenhoeck & Ruprecht, 2006.

———. *Fear, Faith, and the Future: Affirming Christian Hope in the Face of Doomsday Prophecies*. Minneapolis: Augsburg, 1980.

———. *God—The World's Future: Systematic Theology for a Postmodern Era*. Minneapolis: Fortress, 1992.

———. "Happy Danes and Deep Incarnation." *Dialog* 52 (2013) 244–50.

———. *Science, Theology, and Ethics*. Ashgate Science and Religion Series. Aldershot, UK: Ashgate, 2003.

Pihkala, Panu. *Early Ecotheology and Joseph Sittler*. Studies in Religion and the Environment 12. Zurich: LIT, 2017.

Pinches, Charles. "Eco-minded: Faith and Action" (Review of *Earth Community, Earth Ethics*, by Larry L. Rasmussen). *Christian Century* 115/22 (August 12–19, 1998) 755–57.

Presbyterian Church (U.S.A.), Presbyterian Eco-Justice Task Force. *Keeping and Healing the Creation: A Resource Paper*. Louisville: Committee on Social Witness Policy, Presbyterian Church (U.S.A.), 1989.

Rad, Gerhard von. *Genesis: A Commentary*. Translated by John H. Marks. The Old Testament Library. Philadelphia: Westminster, 1961.

Rahner, Karl. *Foundations of Christian Faith: An Introduction to the Idea of Christianity*. Translated by William V. Dych. New York: Seabury, 1978.

Ramsey, George W. "Is Name-Giving an Act of Domination in Genesis 2:23 and Elsewhere?" *Catholic Biblical Quarterly* 50 (1988) 24–35.

Rasmussen, Larry L. *Earth Community, Earth Ethics*. Ecology & Justice. Maryknoll, NY: Orbis, 1997.

———. *Earth-Honoring Faith: Religious Ethics in a New Key*. New York: Oxford University Press, 2013.

———. "Next Journey: Sustainability for Six Billion and More." In *Ethics for a Small Planet: New Horizons on Population, Consumption, and Ecology*, edited by Daniel C. Maguire and Larry L. Rasmussen, 67–140. SUNY Series in Religious Studies. Albany: State University of New York Press, 1998.

Rhoads, David M., ed. *Earth and World: Classic Sermons on Saving the Planet*. New York: Continuum, 2007.

———. "A Theology of Creation: Foundations for an Eco-Reformation." In *Eco-Reformation: Grace and Hope for a Planet in Peril*, edited by Lisa E. Dahill and James B. Martin-Schramm, 1–20. Eugene, OR: Cascade Books, 2016.

Rossing, Barbara R. "The World Is about to Turn: Preaching Apocalyptic Texts for a Planet in Peril." In *Eco-Reformation: Grace and Hope for a Planet in Peril*, edited by Lisa E. Dahill and James B. Martin-Schramm, 140–59. Eugene, OR: Cascade Books, 2016.

Ruether, Rosemary Radford. *New Woman, New Earth: Sexist Ideologies and Human Liberation*. New York: Seabury, 1975

Ruffing, Janet K. "Kataphatic Spirituality." In *The New Westminster Dictionary of Christian Spirituality*, edited by Philip Sheldrake, 117–19. Louisville: Westminster John Knox, 2005.

Russell, Letty M. *The Future of Partnership*. Philadelphia: Westminster, 1979.

———. *Growth in Partnership*. Philadelphia: Westminster, 1981.

Russell, Robert John. *Cosmology from Alpha to Omega: The Creative Mutual Interaction of Theology and Science*. Theology and the Sciences. Minneapolis: Fortress, 2005.

Bibliography

Saler, Robert C. "Joseph Sittler and the Ecological Role of Cultural Critique: A Resource for Eco-Reformation." In *Eco-Reformation: Grace and Hope for a Planet in Peril*, edited by Lisa E. Dahill and James B. Martin-Schramm, 94–109. Eugene, OR: Cascade Books, 2016.

Santmire, H. Paul. "American Lutherans Engage Ecological Theology: The First Chapter, 1962–2012, and Its Legacy." In *Eco-Lutheranism: Lutheran Perspectives on Ecology*, edited by Karla G. Bohmbach and Shauna K. Hannan, 17–54. Minneapolis: Lutheran University Press, 2013.

———. *Before Nature: A Christian Spirituality*. Minneapolis: Fortress, 2014.

———. *Behold the Lillies: Jesus and the Contemplation of Nature—A Primer*. Eugene, OR: Cascade Books, 2017.

———. *Brother Earth: Nature, God, and Ecology in a Time of Crisis*. New York: Nelson, 1970.

———. "Creation and Nature: A Study of the Doctrine of Nature with Special Attention to Karl Barth's Doctrine of Creation." ThD diss., Harvard University, 1966.

———. "Creation and Salvation according to Martin Luther: Creation as the Good and Integral Background." In *Creation and Salvation: Dialogue on Abraham Kuyper's Legacy for Contemporary Ecotheology*. Vol. 1, *A Mosaic of Selected Classic Christian Theologies*, edited by Ernst M. Conradie, 173–202. 2 vols. Studies in Religion and the Environment 5. Zurich: LIT, 2012.

———. "A Critical Challenge for Ecological Theology: Liturgical Renewal." *Sewanee Theological Review* 46 (2003) 423–46.

———. "Ecology, Justice, and Theology." In *Readings in Ecology and Feminist Theology*, edited by Mary Heather MacKinnon and Moni McIntyre, 196–207. Kansas City: Sheed & Ward, 1995.

———. "The Genesis Creation Narratives Revisited: Themes for a Global Age." *Interpretation* 45/4 (1991) 366–79.

———. "Healing the Protestant Mind: Beyond the Theology of Human Dominion." In *After Nature's Revolt: Eco-Justice and Theology*, edited by Dieter T. Hessel, 57–78. Minneapolis: Fortress, 1992.

———. "Images of an Ordinary Conjugal Spirituality." *Spiritus* 20/1 (forthcoming).

———. "Introduction to the Theme." In *Creation: Called to Freedom; Outdoors Ministry Curriculum*, 1–10. Chicago: Division for Congregational Life, Evangelical Lutheran Church in America, 1990.

———. "I-Thou, I-It, I-Ens." *Journal of Religion* 47 (1968) 260–73.

———. *Nature Reborn: The Ecological and Cosmic Promise of Christian Theology*. Theology and the Sciences. Minneapolis: Fortress, 2000.

———. "The Liberation of Nature: Lynn White's Challenge Anew." *Christian Century* 102/18 (May 22, 1985) 530–33.

———. "A New Option in Biblical Interpretation." In *The Travail of Nature: The Ambiguous Ecological Promise of Christian Theology*, 200–218. Philadelphia: Fortress, 1985.

———. "Partnership with Nature according to the Scriptures: Beyond the Theology of Stewardship." *Christian Scholar's Review* 32 (2003) 381–412.

———. "Ecology, Justice, Liturgy." In *Theologians in Their Own Words*, edited by Derek R. Nelson et al., 217–32. Minneapolis: Fortress, 2013.

———. "A Reformation Theology of Nature Transfigured: Joseph Sittler's Invitation to See as Well as to Hear." *Theology Today* 61 (2005) 509–27.

Bibliography

———. *Ritualizing Nature: Renewing Christian Liturgy in a Time of Crisis*. Minneapolis: Fortress, 2008.

———. "So That He Might Fill All Things: Comprehending the Cosmic Love of Christ." *Dialog* 42 (2003) 257–78.

———. *South African Testament: From Personal Encounter to Theological Challenge*. Grand Rapids: Eerdmans, 1987.

———. "The Spirituality of Nature and the Poor: Revisiting the Historic Vision of St Francis." In *Tending the Holy: Spiritual Direction across Traditions*, edited by Norvene Vest, 131–47. Spiritual Directors International Series. Harrisburg, PA: Morehouse, 2003.

———. "Toward a Christology of Nature: Claiming the Legacy of Joseph Sittler and Karl Barth." *Dialog* 34 (1995) 270–80.

———. "Toward a Cosmic Christology: A Kerygmatic Proposal." *Theology and Science* 9 (2011) 287–306.

———. *The Travail of Nature: The Ambiguous Ecological Promise of Christian Theology*. Philadelphia: Fortress, 1985.

———. "The Two Voices of Nature: Further Encounters with the Integrity of Nature." In *Eco-Reformation: Grace and Hope for a Planet in Peril*, edited by Lisa E. Dahill and James B. Martin-Schramm, 71–93. Eugene, OR: Cascade Books, 2016.

Schade, Leah D. *Creation-Crisis Preaching: Ecology, Theology, and the Pulpit*. St. Louis: Chalice, 2015.

Schmid, H. H. "Creation, Righteousness, and Salvation: 'Creation Theology' as the Broad Horizon of Biblical Theology." In *Creation and the Old Testament*, edited by Bernhard W. Anderson, 102–17. Issues in Religion and Theology 6. Philadelphia: Fortress, 1984.

Schreiner, Susan E. *The Theater of His Glory: Nature and the Natural Order in the Thought of John Calvin*. Studies in Theology 3. Durham, NC: Labyrinth, 1991.

Scott, Nathan A., Jr. "The Poetry and Theology of Earth: Reflections on the Testimony of Joseph Sittler and Gerard Manly Hopkins." *Journal of Religion* 54 (1974) 102–18.

Simkins, Ronald A. *Creator & Creation: Nature in the Worldview of Ancient Israel*. Peabody, MA: Hendrickson, 1994.

Sittler, Joseph. *The Anguish of Preaching*. Philadelphia: Fortress, 1966.

———. "Called to Unity." In *Evocations of Grace: The Writings of Joseph Sittler*, edited by Steven Bouma-Prediger and Peter Bakken, 38–50. Grand Rapids: Eerdmans, 2000.

———. *The Doctrine of the Word in the Structure of Lutheran Theology*. Knubel-Miller Foundation Lectures. Philadelphia: Muhlenberg, 1948.

———. "Ecological Commitment as Theological Responsibility." *Zygon* 5 (1970) 172–81.

———. "Ecological Commitment as Theological Responsibility." *Southwestern Journal of Theology* 13/2 (1971) 35–45.

———. "Ecological Commitment as Theological Responsibility." In *Evocations of Grace: The Writings of Joseph Sittler on Ecology, Theology, and Ethics*, edited by Steven Bouma-Prediger and Peter Bakken, 76–86. Grand Rapids: Eerdmans, 2000.

———. *Essays on Nature and Grace*. Philadelphia: Fortress, 1972.

———. *Evocations of Grace: The Writings of Joseph Sittler on Ecology, Theology, and Ethics*. Edited by Steven Bouma-Prediger and Peter Bakken. Grand Rapids: Eerdmans, 2000.

———. *Gravity & Grace: Reflections and Provocations*. Edited by Linda-Marie Delloff. Minneapolis: Augsburg, 1986.

Spencer, Daniel T., and James Martin-Schramm, eds. "Fidelity to Earth: A Festschrift in Honor of Larry Rasmussen." Special issue, *Union Seminary Quarterly Review* 58/1-2 (2004).

Steinmetz, David C. "Luther and Loyola." *Interpretation* 67 (1993) 5-14.

———. *Luther in Context*. Bloomington: Indiana University Press, 1986.

———. "Scripture and the Lord's Supper in Luther's Theology." *Interpretation* 27 (1983) 253-65.

Stendahl, Krister. "The Apostle Paul and the Introspective Conscience of the West." *Harvard Theological Review* 56 (1963) 199-215.

———. "The Apostle Paul and the Introspective Conscience of the West." In *Paul among Jews and Gentiles*, 78-96. Philadelphia: Fortress, 1976.

Stewart, Benjamin A. "The Stream, the Flood, the Spring: The Liturgical Role of Flowing Waters in Eco-Reformation." In *Eco-Reformation: Grace and Hope for a Planet in Peril*, edited by Lisa E. Dahill and James B. Martin-Schramm, 160-76. Eugene, OR: Cascade Books, 2016.

Stewart, Claude Y. *Nature in Grace: A Study in the Theology of Nature*. NABPR Dissertation Series 3. Macon, GA: Mercer University Press, 1983.

TeSelle, Eugene. "How Do We Recognize a *Status Confessionis*?" Theological Table-Talk. *Theology Today* 45 (1988) 71-78.

Tillich, Paul. "Nature and Sacrament." In *The Protestant Era*, 94-111. Translated by James Luther Adams. Chicago: University of Chicago Press, 1948.

Tucker, Mary Evelyn, and John A. Grim. "Introduction: The Emerging Alliance of World Religions and Ecology." *Daedalus* 130/4 (2001) 1-22.

Udall, Stewart L. *The Quiet Crisis*. New York: Holt, Rinehart & Winston, 1963.

Walsh, Sylvia I. "Paradox." In *A New Handbook of Christian Theology*, edited by Donald W. Musser and Joseph L. Price, 346-48. Nashville: Abingdon, 1992.

Westermann, Claus. *Genesis 1–11: A Commentary*. Translated by John J. Scullion. Continental Commentaries. Minneapolis: Augsburg, 1984

Whitelam, Keith. "Israelite Kingship: The Royal Ideology and Its Opponents," in *The World of Ancient Israel: Sociological, Anthropological, and Political Perspectives*, edited by R. E. Clements, 119-39. Cambridge: Cambridge University Press, 1989.

Wilkinson, Loren E. "A Christian Ecology of Death: Biblical Imagery and 'The Ecologic Crisis.'" *Christian Scholar's Review* 5 (1976) 319-38.

———. "Cosmic Christology and the Christian's Role in Creation." *Christian Scholar's Review* 11 (1981) 18-40.

Yeago, David S. "Jesus of Nazareth and Cosmic Redemption: The Relevance of St. Maximus the Confessor." *Modern Theology* 12 (1996) 163-93.

Yeats, William Butler. "The Second Coming." *Poetry Foundation* (website). https://www.poetryfoundation.org/poems/43290/the-second-coming/.

Zachman, Randall C. *Image and Word in the Theology of John Calvin*. Notre Dame, IN: University of Notre Dame Press, 2007.

Index

Adams, Edward, 4n6
Aho, Eric, xvii
Albrecht, Paul, 108n14
Albright school, 108
Althaus, Paul, 45n8, 46n11, 60n59, 68nn82,83, 69n84, 76n103
Anderson, Bernhard W., 4n6, 9n12, 22n40
animals, 16–17, 20–21, 23, 28–31, 33, 35–39, 48, 65, 89, 99, 117, 128
anthropocentrism, 3, 17, 53, 107, 108, 128, 144, 147; see theoanthropocentrism
apocalyptic sensibility, 130
apologetic theology of nature; see theology, apologetic
Aquinas, Thomas, xi, xiv, 44, 73
Auerbach, Eric, 90n20
Augustine, 12, 12n17, 26, 57, 113, 147, 155

Bakken, Peter, 78n2, 106n8
Barth, Karl, xiv, xivn2, xv, 8n12, 9, 52, 66–67, 73, 80, 82, 96, 107n10, 108, 135, 137–40, 143, 145–47, 155
Bauckham, Richard J., 4n6
Benedictine view of nature, 27, 29; see Franciscan view of nature
biblical images of living with nature, 1–41
  awestruck contemplation according to Job, 34–40
  creative intervention according to Genesis 1, 12–26

partnership, 5–6
sensitive care according to Genesis 2, 26–34
Blumenberg, Hans, 47n12
Bonaventure, 147
Bonhoeffer, Dietrich, 92, 93, 113, 114n33, 132, 134–36, 139
Bornkamm, Heinrich, 56n43, 57n45, 59n58, 65n72, 66n75
Boston Industrial Mission, 105n5
Bouma-Prediger, Steven, 41n82, 78n2, 95nn30,33
Braaten, Carl E., 109n15
Brett, Mark G., 19n31
Brown, William P., 10n15, 12nn18,19, 16n24, 17n25, 19–21, 21nn36,37, 22n39, 24, 24n43, 27, 29n54, 32nn57,58,59,60, 33nn62,64, 35, 36, 36nn67,69, 37nn70–73, 38, 38nn74,75, 39, 39nn76,77, 111n22
Brueggeman, Walter, 19n30, 22n38
Brunner, Emil, xiv, xv, xvn4, 66, 108
Buber, Martin, 13, 14, 89, 89n19, 145, 157
Buller, Cornelius, 8n11
Bultmann, Rudolf, 8n12, 80, 91, 108, 137

Calvin, John, xi, 44–45, 47, 50, 59, 62, 72, 72n93, 73, 75, 77, 96, 98, 101, 113, 126, 133, 143, 145, 147, 154
Carroll, James, 142
Carson, Rachel, 138, 138n7

Index

Carter, Vernon, 140
celtic saints, 155
Chidester, David, 47n12
Childs, Brevard, 8n12, 10n15
Childs, James M. Jr., 78n2
Christ
    cosmic, xii, 4, 6, 10, 64, 72–77,
        78–102, 105, 128, 145, 154
    deep incarnation of, 82
    Great Story and, 7, 9, 25, 31, 34,
        39–40
    ubiquity of, 72–76, 101–2
Cobb, John. B., 104n3, 108n13, 116,
    143, 147, 154
Coffin, William Sloane, 142
communicato idiomatum, 75–76
Cone, James, 142, 155
Confessing Church in Germany, 79,
    92, 135–36, 139, 140, 142
Conradie, Ernst M., 125n49
contemptus mundi, 116
coronavirus, xvii; *see* plague
cosmic Fall, 32; *see also* Fall, the
creatio ex nihilo, 12
creation, fulfilment of, 9, 12, 15, 25,
    39, 100
crisis, emergency global ecojustice,
    i, xii, xiii, xvi, 2, 5n7, 15n14,
    41n81, 42, 84, 87, 87n15, 91,
    92, 94, 96, 104n3, 106, 107,
    110, 114n32, 115, 118, 120,
    120n40, 122n45, 123, 124,
    126, 129, 130n54, 138, 138,
    138n7, 143–46, 146n18, 148,
    149, 155
    delineated, 129n53
Cullmann, Oscar, 8n12

Dahill, Lisa, 101n44, 105n5
Daly, Mary, 104n3, 142
Day, Katie, 142
De Gruchy, John W., 113n29
discipleship, 140–42
dominion over nature, human, 7, 19,
    20, 21, 63–64, 84, 111, 144;
    see nature, stewardship of
Dowey, Edward, 44n4

Drummy, Michael F., 107n11, 126n49
Dubos, René, 27, 27n48

earth, the, xvi, 1–3, 4n6, 5n7, 6, 7,
    8n12, 9, 12, 15–17, 19–24,
    26, 27, 27n49, 28, 32, 33, 35,
    35n66, 36, 41, 43, 47, 48,
    61, 63, 65, 66, 81–85, 88, 89,
    95n26, 97, 100, 110n18, 111,
    112, 114n32, 115–19, 122,
    122n46, 128, 130n54, 133,
    154, 155, 157; *see* nature; *see*
    land, the
Easter Vigil, 109n8
Edwards, Denis, 68n81, 104n3,
    125n49
Edwards, Jonathan, 96, 97
Edwards, Mark, 44n5
Eckhart, Meister, 56n44
ecofeminism, 142, 155
ecological reformation of
    Christianity, 105–7, 124, 126,
    157
ecological theology, xii, 82, 90, 98,
    103, 104–10, 113–14, 119,
    122–26, 148, 155–56; *see*
    nature, theology of
ecology and justice, 136–37
ecumenism, 154–56
emergency, global, xv–xvii, 130, 143–
    45; *see* crisis, global ecojustice
Emma, 141
eschatatology, 4, 6, 8, 10–11, 25–26,
    65, 92, 99, 110–13, 124, 127
ethics, xii, 108, 110, 113, 113n28,
    114n33, 115–19, 124, 148,
    156

Fall, the, 9, 23, 25, 31–33, 35–36,
    52, 56, 61–62, 65–66, 71; *see*
    cosmic Fall
finitum capax infinitum, 58, 59, 117
Fletcher, Joseph, 108n14
Fonda, Jane, 142
Fowler, Robert Booth, 112n26
Fox, Maxwell, 148

Index

Francis of Assisi, ix, xi, 27, 99, 113, 146, 147, 155
Francis, Pope, ix, ixn9, x, xi, 41n82, 83, 106, 155
Franciscan view of nature, 27, 29, 65; *see* Benedictine view of nature
Fretheim, Terence, xii, 4n6, 8n11, 9, 10, 12, 14, 17n26, 18, 18n27, 19n32, 30n55, 31n56, 32n61, 33nn63,65, 35n66, 111nn22,24, 124
God; *see* Christ; *see* Luther, God; *see* Spirit
  biblical view of, 6–12
  immanent, 14, 53–63, 73–74, 113, 117, 144
  transcendent, 18, 53–63
  Trinity, 69–71, 152–54
Gottlieb, Roger S., 126n49
Graham, William A., 47n12
Great Story, the, 7, 8, 9n14, 25, 31, 34, 39, 40
Green, Joel B., 7n8, 8n10
Gregersen, Niels Henrik, 62n64, 64n71, 70n90, 71n91, 77n109, 82n7
Grossberg, Daniel, 40n80
Gunton, Colin, 97n39

Habel, Norman, 4n6, 5n7, 10n15, 18n29, 122n44
Hall, Douglas John, 2n3
Hazelton, Roger, 54n35
Hefner, Philip, xii, 109, 109nn17,18, 110, 124
Heinecken, Martin, 139
Hendel, Kurt K., 58n51
Hendrix, Scott H., 44n3
Hessel, Dietrich, 41n82
Hiebert, Theodore, 4n6, 9n12, 20n20, 23, 26n47, 27n50, 28, 28n52, 111n22
Hoffmann, Bengt R., 55n37
Holden Village, 123
Hopkins, Gerard Manley, 94–97, 97n38, 99
Horn, Henry E., 149

Horrell, David G., 4n6
Howells, Edward, 54n36, 56n44
Hunsinger, George, 73, 73n98

iconography, Protestant, 50, 98
images, 17, 50, 93, 98n41, 99n41
Irenaeus of Lyon, 147, 154
I-Thou, I-It, I-Ens relationships, 13, 13n20, 14, 89, 145, 145n14, 157

Jackson, Jesse, 142
Janzsen, J. Gerald, 36n68, 37n73, 39n78
Jenkins, Willis, 103n2
Jenson, Robert W., 7n9, 97n37, 109n15
Job, 10, 11, 13, 23, 34–40, 100, 101
Johnson, Elizabeth, 104n3, 142, 155
Jorgenson, Kiara, 112n26
justice and ecology; *see* ecology and justice
justification by faith, 46, 51, 52, 60n60, 69n85, 92, 127, 156; *see* Luther, justification by faith

Kahl, Brigitte, 28n52
Kant, Immanuel, 79, 137
Kaufman, Gordon D., 13, 107, 107n9, 138, 146
Kleckley, Russell, 45n8, 46n10, 53n31, 58, 59, 59nn56,57, 72, 72n95, 73
Knierim, Rolf P., 4n6, 32n61
Koerner, Joseph Leo, 50n22

land, the, 16, 20, 21, 22n39, 28–30, 33, 137; *see* earth; *see* nature
Landes, George, 9n12
Lane, Beldon C., 50n24
Lathrop, Gordon, xii, 112n25, 124, 151
Levenson, John D., 22n41, 88n17, 90n20
Levin, David Michael, 47n12

INDEX

liberation, 9n10, 34, 116, 134, 142, 143, 143n9, 150, 155
Lindberg, Carter, 64n68
liturgy, xvi, 93, 98, 112, 122, 149–53; *see* ritual
Loefgren, David, 45n8
Lohse, Bernhard, 44, 44nn6,7, 45, 45n7, 51n27, 52nn28,30, 58, 69n83, 72n94, 76n104
Loyola, Ignatius of, 51, 51n25, 76n104
Luther, Martin, x–xi, xiii, 1, 12, 42–47, 47n14, 48n15, 49nn17–20, 50nn21,23, 51, 51nn26,27, 52nn29,30, 54n34, 55nn38–42, 56n43, 57nn45,48, 58, 60nn59,60, 59nn55,58, 61n61, 62nn62,63, 63nn65,66,67, 64nn69,70,71, 65nn72,73,74, 66nn75,76, 67n79, 68nn80–83, 69n84, 70nn86–89, 72, 73nn96–98, 76, 76nn103,104, 77, 78–80, 86, 91–94, 97–98, 98n41, 100, 101n48, 102, 105n6, 113, 114n33, 117, 118, 118n38, 119, 126, 128, 128n51, 130–32, 134, 138–40, 143, 145, 145n17, 147, 149–50, 155, 157
  cosmic Christ, 72–77
  creation and salvation, 44–45
  Divine curse of nature; *see* nature, Divine curse of
  Divine immanence and transcendence, 53–60, 74
  duplex life of faith with nature, 60–66
  fatherly heart of God, 63, 69–71
  finitum capax infinitum, 58–59
  God, 53–60
  human body, 58, 62, 74
  human dominion as service, 63–64
  incarnation of Christ, 47, 68, 76, 82, 88, 96

justification by faith, 46, 51–52, 60, 69, 126–27
  mysticism, 54–55, 74
  nature as omnimiraculous, 59
  omnipresence of God, 55–60
  paradox, 53–60
  sensibilities of hearing and seeing, 47–51
  theology of the cross, 51–52
  theology of glory, 51–52
  theology of the Word, 46–50
Lutheran maximalism, 126–30; *see* Lutheran minimalism
Lutheran minimalism, 126–30; *see* Lutheran maximalism
Lutherans Restoring Creation, xvii, 121, 123
Lutz, Paul E., 120n42, 142

Mackinnon, Mary Heather, 136n4
macro-narrative, biblical, 10; *see* Great Story
McDaniel, Jay B., 15n23
McFague, Sally, 53, 54n33, 56, 75n102, 104n3, 142, 155
McKibben, Bill, 129n53
metaphorical reductionism, 36, 57
micro-narrative, 10; *see* Great Story
Miles, Margaret, 49n17, 96n36
Miller, Perry, 133, 144, 144n13
miracles, 59, 128; *see* nature, omnimiraculous view of
Moe-Lobeda, Cynthia D., 114n33, 125n48
Moltmann, Jürgen, 4n6, 10, 17, 53, 54n33, 56, 77, 78n2, 97n34, 106, 154, 155
Morgan, David, 98n41
Muir, John, 101, 132–134, 136, 145
Mullaney, Anthony, 142
Murphy, George L., 77n107, 122n44, 128n51
Mysticism, baptismal, 153; *see* Luther, mysticism

Nash, James, 41, 104n3, 105n5, 146, 154

INDEX

natural theology, 79, 96–97, 108
naturalism, 81–82
nature; *see* earth; *see* land, the
   apologetic theology of; *see*
      theology, apologetic
   as kin, brother, sister, 39, 82, 84,
      121, 144
   awestruck contemplation of, 6,
      23, 34, 35, 39, 40n80, 41
   church activism in behalf of,
      119–23
   creative intervention in, 6, 20,
      26–34
   dark side of, vii, 23, 34–39, 100;
      see nature, Divine curse of
   definition of, x, 43
   Divine curse of, 20, 27, 31–35,
      61–62, 66
   integrity of, xiv, 6, 14–15, 17,
      21–22, 27–28, 71, 83, 85, 95,
      115, 120, 129, 144
   groaning of, 100, 128, 137
   liberation of; *see* liberation
   living with, according to the Bible
      1–41
   omnimiraculous view of, 59
   praising God, 8, 14, 33, 36, 58, 82
   reconstructionist theology of; *see*
      theology, reconstructionist
   revisionist theology of; *see*
      theology, revisionist
   ritualized, 3, 73, 77, 98, 112, 149
   sensitive care for, 6, 11, 27–30,
      34–36, 40–41, 84–85, 101,
      111, 117, 122
   sexuality and, 40n80
   spirituality of, 153–154
   stewardship of; *see* stewardship
   theology of, xii, xv, 4, 4n6, 5n7,
      9n12, 12, 14n22, 35, 43,
      106–7, 144–49; *see* ecological
      theology
   voices of, 14, 100

Oberman, Heiko, 44n3, 139
Origen, 147
Orr, David, 129n53

outdoor ministries, xii, 123, 146

panentheism, 67, 75
Pannenberg, Wolfhardt, 8
pantheism, 59–60
paradox, 53–60, 74–75, 127, 156
partnership, xi, 1–2, 4–6, 11, 16,
   20–21, 26–27, 35, 38–41; *see*
   stewardship
Pelikan, Jaroslav, 46n10, 47n14,
   50n23, 68n80
Pesch, Otto, 67n78
Peters, Ted, xii, 76, 82n7, 110,
   110nn19,20, 111, 111n21, 124
Pihkala, Panu, 78n2
Picasso, Pablo, 94–95
Pinches, Charles, 114n32
plague, xvii, 33, 61
preaching, 50, 94–95, 107, 121–22,
   148–49, 151–53
Priestly Writers, 10–22, 22–31,
   33–35, 40

racism, 104n3, 137, 140, 141
Rahner, Karl, 68n81
Ramsey, George W., 29n53
Rasmussen, Larry, xii, 41, 113,
   113nn28,29, 114nn32,33, 115,
   115nn35,36, 116–17, 117n38,
   118, 119, 123, 124, 125,
   129n52, 151
reconstructionist theology of nature;
   see nature, theology of,
   revisionist
revisionist theology of nature;
   see nature, theology of,
   reconstructionist
Rhoads, David, xvii, 105, 121, 122,
   122n45, 148
ritual, 109, 112, 148–52; *see* liturgy
ritualizing nature; *see* nature,
   ritualized
Rossing, Barbara R., 149n23
Ruether, Rosmarie Radford, 104n3,
   108, 142, 155
Ruffing, Janet K., 54n36
Russell, Bertrand, 14

Index

Russell, Letty M., 5n7
Russell, Robert John, 77n107

Saler, Robert C., 78n2
Sallman, Warner, 98n41
Santmire, H. Paul, xiv, xivn3, 1n1, 3n5, 4, 4n6, 5, 9nn12,13, 10n16, 12, 13, 13n20, 14n25, 15n23, 26n46, 32n61, 33, 35n66, 40n79, 41n81, 42n1, 45n9, 53n32, 57, 58, 66n77, 72n92, 73n99, 77nn105,108, 78n1, 79n2, 81n6, 97n39, 98n40, 99n42, 100n43, 105n4, 106, 107n10, 108, 109n15, 112, 112n26, 120n41, 123n47, 126n50, 128n51, 131–57
Schade, Leah D., 122n45
Schmid, H. H., 9n12, 25n45
Schreiner, Susan E., 50n24, 101n45, 126n49
Scott, Nathan, 95n26
Second Vatican Council, x, 152, 155
sensibility of hearing, 47–52, 64–65, 67, 80, 91–93, 99; *see* sensibility of seeing
sensibility of seeing, 49–52, 57, 64–65, 67, 71, 77, 80–81, 86–90, 91–93; *see* sensibility of hearing
sexuality; *see* nature, sexuality
Simkins, Ronald A., 4n6
Sittler, Joseph, xii, xiii, xv, 6, 14n23, 15n23, 49, 78, 78n2, 79, 80, 80n4, 81, 81n5, 82–86, 86nn9,10,11, 87nn12,13,14,16, 88, 89n18, 90–93, 93nn21–23, 94, 94nn24,25, 95, 95nn27,28,29,31,32, 96, 96nn34,35, 97, 97n37, 98, 98n41, 99, 101, 102, 105, 106n7, 107, 107n12, 108, 114, 114nn32,33, 115, 119, 120n41, 124,nn37,39, 128, 131, 145, 146, 146n17, 149, 149n23, 156, 157

emergence as a theologian of nature, 80–86
legacy, 97–102
sensibility of seeing, the, 86–92
witness of Picasso and Hopkins, the, 93–96
Spirit, 1, 9, 15, 24, 41, 60, 70–71, 73, 100, 109, 111, 126–28, 151, 153–54
spirituality, 54, 56, 77, 93, 96, 99, 101, 109, 127, 132, 146, 148–49, 153–54; *see* Trinity Prayer
status confessionis, xvi, 130
Steinmetz, David C., 42n2, 49, 51, 51n25, 54n34, 55n42, 65n73, 73n96, 76n104
Stendahl, Krister, 138, 143
stewardship, xi, 2–6, 21, 82–83, 116, 120–21, 131, 148; *see* partnership
Stewart, Benjamin, 98n40
Stewart, Claude Y., 112n26, 143

TeSelle, Eugene, 130n54
Teilhard de Chardin, Pierre, 13n21, 14, 74, 143, 147–48, 152, 155
theoanthropocentrism, xiv–xv, 8–9, 52, 79, 82, 107–9, 112, 126–27, 137–43; *see* theocosmocentrism.
theocosmocentrism xiii–xviii, 5, 82–83, 85, 108–15, 121, 124, 126, 128, 156; *see* theoanthropocentrism.
theology; *see* ecological theology
apologetic, 138
objects of reflection by, xvii–xv
reconstructionist, 81, 107–8, 148
revisionist, 80–81, 108, 148
Thoreau, Henry David, 101n44, 123, 132–34, 136, 144
Tillich, Paul, 11, 54, 105n6, 107n11, 126n49, 132–36, 138–40, 144, 151
Trinity Prayer, 154
Tucker, Mary Ellen, 104n3

Index

Udall, Stewart, 138, 138n7

Van Gogh, Vincent, 69, 98
Von Rad, Gerhardt, 8n12, 25, 25n44

Walden Pond, 134
Walsh, Sylvia I., 54n35
Westermann, Claus, 30n55
White, Jr., Lynn, 113, 143, 146–47
white privilege, 141; *see* racism
Whitelam, Keith, 18n28
wilderness, the, 36, 38–39, 59, 101, 123, 128, 133, 136–37, 144, 154
wildness, 34–36, 100–101
Wilkinson, Loren E., 10n15, 35n66

worship; *see* liturgy
Wright, G. Ernest, 8n12, 108, 111, 137

Yahwist, the 10, 11, 13, 16, 20, 23, 26–34, 35, 38, 40, 111
Yeago, David S., 74nn100,101, 77
Yeats, William Butler, 97n38
Young, Andrew, 142

Zachman, Randall, 47n12, 50n24
Zwingli, Huldrych, 58, 72, 128